New Atlantis
and
The Great
Instauration

Crofts Classics

GENERAL EDITOR

Samuel H. Beer, *Harvard University*

FRANCIS BACON

New Atlantis
and
The Great
Instauration
REVISED EDITION

EDITED BY

Jerry Weinberger

MICHIGAN STATE UNIVERSITY

Harlan Davidson, Inc.
Wheeling, Illinois 60090-6000

Visit us on the World Wide Web at www.harlandavison.com

Library of Congress Cataloging-in-Publication Data

Bacon, Francis, 1561–1626.
   [New Atlantis]
   New Atlantis ; and The Great Instauration / Francis Bacon; edited by
Jerry Weinberger.—2d ed.
   p. cm.—(Crofts classics)
   Second work, translation of Instaurio magna.
   ISBN 0-88295-126-2
   1. Philosophy—Early works to 1800. 2. Philisophy of nature—Early
works to 1800. 3. Science—Philosophy—early works to 1800. 4. Utopias—
Early works to 1800.
I. Weinberger, Jerry.   II. Bacon, Francis, 1561–1626, Instauratio
Magna. English, 1989.   III. Title.   IV. Title: New Atlantis.
V. Title: The Great Instauration
BF1195.N48   1989   100-dc.19   88-25789

contents

introduction
to the revised edition

Along with Machiavelli, Hobbes, and Descartes, Francis Bacon was one of the founders of modern thought. These founders coupled realistic politics with a new science of nature in order to transform the age-old view of mankind's place in the world. They argued that once the efforts of the human intellect were directed from traditional concerns to new ones—from contemplation to action, from the account of what men ought to do to what they actually want to do, and from metaphysics to the scientific method for examining natural causes—the harsh inconveniences of nature and political life would be relieved or overcome. No longer to be revered or endured, the worlds of nature and society would become the objects of human control.

Bacon called his enterprise "the great instauration," an ambiguous term that means at once great restoration and great founding. But he left no doubt that he was engaged in something altogether new: His restoration—his reform of the ways and means of human reason—would in fact be a founding because its aim would be "to lay the foundation, not of any sect or doctrine, but of human utility and power," in order to "conquer nature in action."[1] Bacon argued that before his instauration reason had suffered at the hands of the ancients, especially Plato and Aristotle. Their dogmatic preference for contemplation over action betrayed contempt for the practical arts, a contempt much more harmful than noble. For it merely served to hide the real courses of nature from view, so that from Aristotle one hears "the voice of dialectics more often than the voice of nature" and

[1]*Below,* pp. 16, 21.

in Plato one sees that "he infected and corrupted natural studies by his theology as much as Aristotle did by his dialectic."[2] Thus a mere prejudice caused the ancients to spend their energies wrangling about the meaning of nature as a part of the cosmos, and about our rightful place in the whole and before the gods, instead of discovering how nature's motions and processes can be bent to human purposes. Human life needs tools for action, but from the ancient wisdom we get nothing but theological and metaphysical speculation. Had they been confined to their own academic business, the ancient teachings would have been merely useless. But reason, however contemplative, cannot but affect practical life, and so the ancients were tempted to concern themselves with politics, especially after Socrates, who was famous for having brought philosophy down from the heavens. Approached from the contemplative and speculative points of view, however, the ancients' concern for practical affairs was in fact impractical and served merely to fuel sectarian controversies about justice and religion, controversies that are inevitable when we are faced by material scarcity and a cosmos that is hostile to our wills and indifferent to our needs.

For Bacon, reason directed by new means and ends would change completely man's place in the whole of things. When properly empowered, reason would no longer merely soothe our enslavement to a stingy nature, or take sides in the political disputes caused by nature's penury. Rather, when governed by the "art itself of interpreting nature," reason would enable man to "conquer nature in action," which is the "real business and fortune of the human race." This real business and fortune would make human will the measure of nature and would not blush at taking aim even at the corruptibility and mortality of the human body, the conquest of which Bacon called the "noblest work" of natural philosophy.[3]

[2] *"The* Refutation of Philosophies" in *The Philosophy of Francis Bacon*, ed. Benjamin Farrington (Liverpool: Liverpool University Press, 1964), pp. 112–15.

[3] *De Sapientia Veterum* XI, in *The Works of Francis Bacon*, ed. James Spedding, Robert Leslie Ellis, and Douglas Denon Heath, 14 vols. (London: Longman and Co., etc. 1857–74), hereafter BW, VI, 646, 721; below, pp. 21, 31, 33.

Bacon's project was thus far more than a mere reformation of human reason and even more than the founding of a new intellectual or political order. For every founding prior to his own had been limited by something beyond the powers of the founder, whether it be nature, fortune, or God. With Bacon's project we encounter an altogether new horizon, one that forces a reconsideration of the words: "In the beginning . . ."

the great instauration

It is fashionable now to have doubts about modern science and to acknowledge that its blessings can be mixed. But even so, we entertain such thoughts from within a world already transformed by modern science. Bacon's thought about such a world is thus important because he could take nothing of it for granted. For Bacon the differences and connections between prescientific and postscientific experience were much closer at hand—though no more important—than they are for us. Bacon knew more immediately than do we that the meaning of natural science can be grasped only by considering questions now assumed to be outside the scope of natural science. He did not regard the new science of nature to be an autonomous enterprise with no links to other human concerns. Moreover, he knew that the scientific transformation of the world would have extraordinary moral and political consequences and that it would pose new problems in place of old ones: In a world to be conquered rather than endured, what moral and political principles will guide the human energies released by the activity of conquest? And in a world freed from God's providence—a world in which we literally choose our own destiny—what principles will define the objects of our choice? And what are the best means for overcoming the resistance of those who would stand to lose with the passing of the old world, or of those too timid to welcome wholeheartedly a new and unknown one? Bacon's teaching about modern science is ultimately about human relations during and after the conquest of nature. Thus the core of his teaching about science is to be found not in his accounts of scientific method, but rather in his political science.

But since the world promised by Bacon's instauration is radically new—so new that it requires of men that they give up much that is familiar and dear—and since it is not wise to discuss in the presence of the resistant the means for disarming them, Bacon made the most audacious aspects of his political teaching difficult to see. The first task for the student is thus to see clearly how Bacon underscores the importance of his political teaching to his project for human reason, an importance by no means obvious. In following Bacon's clues, one learns how to penetrate the teaching to which they lead. The reader thus receives important directions for discovering Bacon's political science, which is an essential source for understanding the meaning of our modern scientific age.

Of the works that state the aim of Bacon's instauration, one stands out from the rest because it presents an explicit organization of Bacon's various writings according to the guiding principle of his scientific project. The short work entitled *The Great Instauration*, reprinted in this volume, was published in 1620, only six years before Bacon's death, as an introduction to his treatise on scientific method, the *Novum Organum*. Originally published in Latin, the several pages of *The Great Instauration* describe formally the separate parts of Bacon's project. Since he called his project the "great instauration," *The Great Instauration* is the title of the work outlining the plan of the "great instauration," which would be a series of writings detailing the geography of the new world of reason. Thus *The Great Instauration* explains that the first part of the instauration would be an account of which sciences are complete and which need work or have not been attended to at all, the second part would be an account of proper scientific method, the third would present a natural history, or an account of the phenomena to be explained by science, the fourth part would set forth models of how his new method would be applied to particular subjects, the fifth would set forth some of Bacon's discoveries from the use of traditional methods, and the sixth part would set forth the completed philosophy according to his overall plan.

Now a glance at Bacon's whole corpus shows that most of the parts outlined in *The Great Instauration* were not completed or else seem to be missing altogether.[4] And the most important one, the sixth part that is to disclose and set forth the "developed and established" new philosophy that will treat the "real business and fortunes of the human race, and all power of operation," the part to which the rest of the instauration is "subservient and ministrant," is said by Bacon himself to have been beyond his strength and hope to finish.[5] But we must not take Bacon's modesty entirely at face value. For if the sixth part were to exist, it would describe the ends of science, where "human knowledge and human power do really meet in one"; and of such a picture Bacon says not that it is beyond his powers and so impossible for him to provide but rather that it "cannot easily be conceived or imagined."[6] This means, of course, that with difficulty such a picture can be conceived or imagined. In other words, if the sixth part of the great instauration describes the final goal of science, then presenting that part is not beyond Bacon's strength and hope, although it will be difficult for readers to see. And the *New Atlantis* presents just such a picture, for it depicts a society dedicated to the pursuit of science and the conquest of

[4]*Bacon* says that the first part is wanting, although "some account of [it] will be found in the second Book of the 'Proficience and Advancement of Learning'." (See below, pp. 19–20, 33.) The new method (the second part) was provided in the *Novum Organum*, but, according to Bacon, only as a "summary digested into aphorisms." (See below, pp. 20–26, 33.) The natural history is the most filled out, although it too is incomplete as measured by Bacon's description of it in *The Great Instauration*. (See below, pp. 26–29, and F.H. Anderson, *The Philosophy of Francis Bacon* [Chicago, 1948], pp. 34–36, 40.) Part four is missing except for a very small fragment. (See below, p. 30; *Scala intellectus sive filum labyrinthi*, BW, II, 687–89.) Part five is likewise missing except for a small fragment. (See below, pp. 30–31; *Prodromi sive anticipationes philosophiae secundae*, BW, II, 690–92.) And there is no Baconian work identified as part six.

[5]*Below,* pp. 26, 31.

[6]*Below,* p. 31.

nature. Bacon names this society "Bensalem," which means "offspring of peace, perfection, or wholeness." Bensalem is, moreover, dedicated to ends that are "compounded of all goodness." But why, then, did Bacon not simply say that the *New Atlantis* was his account of the ends of the new project for reason, that it was the sixth part of the instauration, and that it was not beyond his powers?[7]

The reason seems to be that Bacon was right in the first place. For by long tradition the *New Atlantis* has been thought to be incomplete. According to Rawley, Bacon's secretary who oversaw the posthumous publication of the *New Atlantis*, the work was left unfinished because Bacon was deterred by considerations of time and preference from composing the "frame of Laws, or of the best state or mould of a commonwealth," the account of scientific politics, that would have completed it.[8] But before we accept Rawley's opinion, it is important to remember Bacon's claim that the sixth part of his project would be difficult to conceive or imagine. In *The Great Instauration* Bacon explains this difficulty as the result of men "thinking the matter in hand" to be "mere felicity of speculation" rather than the mastery of nature.[9] But in his division of the sciences in the *Advancement of Learning*, Bacon provides an additional explanation, at least as regards politics. There he says that political science, being dangerous, is a subject "secret and retired."[10] If the end of science is difficult to imagine, this may well be because the account of its politics must be "secret and retired." Therefore, the *New Atlantis* may only *appear* to be incomplete and thus impossible to grasp in its entirety. We must consider that the *New Atlantis* may indeed be the picture of the end

[7]*Below*, pp. 46, 60. For a fuller account of the place of the *New Atlantis* in the whole of Bacon's corpus, see Jerry Weinberger, "Science and Rule in Bacon's Utopia: An Introduction to the Reading of the *New Atlantis*," *American Political Science Review* (1976):866–72.
[8]*Below*, pp. 36, 83.
[9]*Below*, p. 31.
[10]*BW*, III, 473–76.

of science as anticipated in the sixth part of the plan, and that it is a *complete* picture that appears to be incomplete because it presents a "secret and retired" teaching about politics that can be discerned only with difficulty.

The Great Instauration tells us that the sixth part of Bacon's project is the crucial center of his account of science and human destiny. So long as this part is complete and discloses fully the true ends of scientific progress, the other parts need not be finished by Bacon himself. They can be the work of others to follow. But *The Great Instauration* and the *New Atlantis* force us to wonder about the crucial sixth part: It may be complete in the sense that it presents a dangerous, and therefore secret and retired, political science. But does the danger lie on the way to the end of science, or in that very end itself? Does Bacon worry only about those who fear losing the old world, or does he warn us about the new one as well? Bacon's intentions are always bolder than he openly admits them to be. And thus in approaching his teaching we are forced to be slow and cautious. The reader must be prepared to study the *New Atlantis* with great care, paying attention to Bacon's silence in addition to his speeches and watching for the complex fabric of allusions that forms the structure of the story. The full importance of Bacon's teaching about the new science can be seen only when the reader realizes that the center of that teaching is a "secret and retired" political science.

the new atlantis

The title and Rawley's postscript declaring the incompleteness of the tale are the most obtrusive parts of the *New Atlantis*. It is far from obvious why "New Atlantis" should even be the title of the work. "Bensalem," "Renfusa," or "Salomon's House" would all be more plausible titles, for Atlantis is mentioned only as the presumably kindred neighbor (later destroyed by natural catastrophe) of the country defeated by the ancient Bensalemites. The title thus calls our attention to another important account of Atlantis, to which Bacon refers without actually naming the author. That author is, of course, Plato, who tells the story of Atlantis in his dialogues *Timaeus* and *Critias*.[11] Now like the *New Atlantis*, Plato's story is incomplete. For the dialogue breaks off just as Zeus, angered by the prideful insolence of the technically proficient Atlantis, prepares to announce his intended chastisement of Atlantis to the assembly of the gods: we do not hear about the moderating that was to produce more harmony in the Atlantians.[12] In apparent confirmation of Rawley's opinion, Bacon's story is like Plato's not only in its subject, that of a technically advanced island, but also is similar in its incompleteness: for like Plato's story, Bacon's seems to be missing a speech about politics, a speech that would doubtless have been similar to Zeus's in considering the themes of punishment and harmony, pride and moderation.

And yet in another respect the *New Atlantis* is indeed finished by comparison to Plato's *Critias*. For whereas the speech of Zeus describing the defects and future perfection of Atlantis is omitted, the speech of the Father of Salomon's House describing the "true state" of Salomon's House, the "noblest foundation" ever upon the earth and the "lanthorn" of Bensalem, is provided in order to reveal the perfection of the new Atlantis "for the good of other nations."[13] Bacon replaces a missing speech about the moderation of excess with a speech about the structure of Salomon's House, the engine of the scientific conquest of nature, and this scientific speech makes no mention of moderation and excess. Yet the Father concludes his speech (and the *New Atlantis*) by giving the narrator a massive tip or "largess," despite the fact that throughout the story the Bensalemite's scorn of tipping has been a mark of their apparent moderation and restraint.[14]

The difference between Plato's and Bacon's stories is that Bacon's Bensalem, unlike Plato's Atlantis, has per-

---

[11]*Timaeus* 19b–d, 22a, 25a–b6; *Critias* 120d6–121c4; cf. 106a–b8. See *Phaedrus* 278d7; *Laws* 676ff.

[12]*Critias* 120d6–121c4.

[13]*Below*, pp. 71, 58, 83.

[14]*Below*, pp. 39, 41, 43, 46; cf. 59, 61, 62, 83.

fected the science that protects her from external harm and nature's corruptibility, indeed from any divine revenge intended by the likes of Zeus. But the two stories are alike in the omission of speeches about excess and moderation. Are we to assume that Bensalemite science by itself assures that the Bensalemites will be harmonious and moderate? Or does the final deed of the story suggest that, despite appearances to the contrary, Bensalem is neither harmonious nor moderate because it lacks the knowledge to be gained from hearkening to Zeus? Perhaps, then, Rawley was both mistaken and yet correct about the completeness of the *New Atlantis*: it may be complete because it contains a secret teaching about scientific politics, but it may also be incomplete because that teaching provides no effective practical wisdom about harmony and punishment, or about pride and moderation.

There are two general themes in the *New Atlantis*. The first theme presents in two parts what happens to the sailors who visit Bensalem: we see their plight while they think themselves subject to necessity and chance, and we see their experiences after they come to regard themselves as free men. The second general theme presents in four parts the revelation of Bensalem to the sailors: we see their reception by the Bensalemites, we hear a history of Bensalem, we hear about Bensalemite erotic behavior, and we hear a description of the "House of Salomon."[15] The two general themes and their parts overlap; how they fit together shows what the promise of Bensalem means for the sailors' liberation.

At first glance it seems that in happening upon Bensalem the sailors are freed immediately from dangers beyond their control. But this first impression fades with closer inspection. First, it is likely the Bensalemites, who have control over wind and sea, have themselves forced the sailors to their shores for purposes of their own.

[15]*The* two sections of the first major theme are: 1) below, pp. 37–60; 2) below, pp. 60–83. The four sections of the second major theme are: 1) below, pp. 37–46; 2) below, pp. 46–60; 3) below, pp. 60–68; 4) below, pp. 68–83.

Second, after their initial confinement the sailors are informed by the Bensalemites that they—the sailors—may be told some things they "will not be unwilling to hear," which of course implies that there are other things the sailors would not have been willing to hear.[16] And finally, the narrator suggests that their initial confinement may be an ominous surveillance and not the harmless custom the Bensalemite official says it is. The narrator's suspicion is confirmed by facts later revealed: according to the priest who is the governor of the Strangers' House, the sailors may anticipate a long and enjoyable stay because the House is well-stocked, it having been thirty-seven years since any strangers had visited the island. In the same vein the priest later remarks—while explaining the Bensalemites' "laws of secrecy" restricting travel in and out of the island—that no strangers have been detained against their will and that the strangers must think that anything reported abroad by returning strangers would be "taken where they came but for a dream." But the lawgiver, Solamona, who promulgated the laws of secrecy, ordained the kindly treatment of strangers permitted on the island precisely because it was "against policy" that strangers "should return and discover [reveal] their knowledge" of Bensalem.[17] That is, the Bensalemite laws concerning strangers and travel, intended to protect the island from foreign corruption, presume the credulousness of non-Bensalemite peoples, contrary to the priest's soothing remarks. The narrator's fears were not unwarranted. For if Bensalemite law and policy are consistent, strangers unwilling to stay, or those judged unfit to stay, must have been restrained by force or killed. This doubtless the sailors would not have been willing to hear.

Thus the Bensalemites' hospitality and humanity are at the very least ambiguous. And since the sailors represent those who will bring Bensalemite science to the rest of humanity, Bacon makes us consider that the means to scientific happiness—the very courses of progress—may be less than benign. Scientific progress may pose

[16]*Below*, p. 44–45.
[17]*Below*, pp. 46, 50–51, 55–59.

the problem of immoral means used for moral ends. But the even more important question concerns the ends of scientific progress themselves. Are there secrets about Bensalem that remain after the sailors' "discovery" of Bensalem to the rest of the world? If so, then the *New Atlantis* may be incomplete after all, and this because of problems inherent in the world's scientific destiny.

The history of Bensalem, related by the priest, answers the sailors' questions about the island's conversion to Christianity and about why the island, at least some of whose inhabitants know all about the outside world, is unknown to that world, a condition caused by the Bensalemite policy concerning the admission of strangers and by the laws of secrecy governing the few Bensalemites who travel abroad. Bensalem's conversion to Christianity appears to be the result of a miracle. But again the appearance fades with closer inspection. For the miracle is confirmed to be genuine by one of the "wise men" of Salomon's House—that is, by a scientist—who distinguishes between miracles, natural occurrences, and "impostures and illusions of all sorts."[18] In other words, the wise man verifies the miracle by certifying rationally that it has no natural or artificial cause. But in the Bible there is no such instance of verification, the miracles of God needing no rational explanation. It is useful in this context to refer to the treatment of miracles in the *Advancement of Learning*, where Bacon makes it clear that there is no standard for distinguishing between miracles and "prodigious narrations" and natural "rarities"—that is, between miracles and natural accidents—because all are by definition rare. In the final analysis there can be no certain distinction between miracles and natural accidents, with the result that there is no certain standard for testing the veracity of those who report and interpret extraordinary events as the signs of divine will. In the matter of miracles there is always room for art, fraud, and illusion. And in this case the interpreter is a scientist of Salomon's House, where, we learn later on, all the

[18]*Below*, pp. 47–49.

techniques of art and illusion have been mastered.[19] Moreover, this testimony, which makes us doubt miracles as such, concerns an instance of original revelation (to a crowd, not an isolated individual). But without trustworthy revelation, guidance by divine will or law is impossible. If Bensalem frees the sailors from natural necessity but not from other dangers, this may result from the Bensalemites' freedom from divine law and revenge, however much the contrary may appear from the surface of the story.

The priest's speeches in the sequel present the civil history of Bensalem, in order to explain to the sailors something of why the extraordinary island is unknown to the rest of the world. In fact, the history recounts the coming of science and scientific rule to Bensalem. And the most impressive aspect of this history, not even discussed in the story, is that for Bensalem, unlike the rest of the world, the advent of science preceded the coming of Christianity. We know this from the obvious fact that a wise man of Salomon's House, the scientific establishment founded by King Solamona, verified the miracle revealing Christianity to the island, and from the less obvious but still clear chronological evidence from the story: If the "six score years" since navigation had begun to increase, mentioned by the priest, be reckoned from the obvious beginning of European navigation, 1492, the sailors' voyage took place in 1612.[20] Therefore, the time referred to by the priest as "three thousand years ago," when the great Atlantis "did flourish," would be

[19]*Below,* pp. 77–78. *Advancement of Learning,* hereafter *Advancement,* BW, III, 287–88. See 333–37, 340–42; cf. *De Augmentis,* BW, IV, 243–44.

[20]*See* Howard B. White, *Peace Among the Willows: The Political Philosophy of Francis Bacon* (The Hague: Martinus Nijhoff, 1968), pp. 104 n. 30, 121–22. The 1492 date puts the voyage after the publication of the *Advancement of Learning* and before the *Novum Organum* and *De Augmentis,* and it provides a clue to the primacy of the *Advancement of Learning* in the whole of Bacon's writings. The division of the sciences contained in the *Advancement of Learning* provides the description of the intellectual globe that will tame and conquer the terrestrial globe. Cf. White, pp. 104 n. 30, 121–22, 197–98.

1388 B.C.E., 458 years before the biblical Solomon completed the temple in 930 B.C.E., and the time of King
Solamona's rule in Bensalem, "about nineteen hundred
years ago," would be 288 B.C.E.[21] In other words, the
coming of science to Bensalem preceded the coming of
Jesus to the world by almost three hundred years. If we
are forced by the story to ask why the Christian world
needs Bensalemite science, we are equally compelled to
wonder why science, as it appears in Bensalem, needs
Christianity—a Christianity quite possibly managed more
by art and illusion than by sincere faith.

We can follow up these questions by considering the
Advancement of Learning, where Bacon has some interesting things to say about the order of historical destiny.
There he says that in the modern age men will imitate
the heavens by circumnavigating the globe "as the heavenly bodies do" and that God has ordained that such
circumnavigation will be coeval with "the further proficience and augmentation of all sciences."[22] Now the
Bensalemites engage in such circumnavigation, and the
narrator, to whom the wonders of Bensalemite science
have been revealed, writes in English to an English au-
dience, as if Great Britain were to lead the world to its
Bensalemite destiny. The British—*the* modern seafaring
people—are in fact the people who will become Bensalem when guided by Bacon's instauration. No wonder,
then, that the crucial period of British history, the period
from "uniting of the Roses to the uniting of the Kingdoms" (from Henry VII's founding of the modern British
state to the combining of the English and Scottish mon-
archies—one hundred years before actual political union—
with the accession of James I), is "after the manner of
the sacred history, which draweth down the story of the
Ten Tribes and of the Two Tribes as twins together."[23]
For in this age the modern state was built and Bensalem—
modern science—was revealed to it by Bacon's great instauration. But if Britain is now joined in monarchy "for
ages to come," and if British and Bensalemite destiny

[21]*Below*, pp. 51–52, 56.
[22]*Advancement*, BW, III, 340.
[23]*Ibid.*, p. 336.

are the same, ushering in a wholly new age for mankind, it is surprising that in mentioning the likeness of British history to the sacred history Bacon is silent about the coming of Jesus: he speaks only of the sacred history before the coming of Jesus, in particular the sacred history from the Exodus to the division of David's empire.[24] Not only in Bensalem, then, but in the world that would receive Bensalemite science, the true order of things is for Christianity to appear after science. But this is surely perplexing, unless we assume that science is not really different from Christianity—that science is but a mode of Christian faith. Only with such an assumption does Bacon's story not reverse the actual historical order. And with this assumption the Christianity introduced to the Bensalemites would have been but a new version of their old and established science, which had been all along a form of charitable belief.

But then what are we to make of the renewal of science by Christianity? For not only is that Christianity in essence scientific, but as scientific it has no need for Jesus. According to the clues about historical destiny provided by the *Advancement of Learning*, the true beginning of the new age—the concatenation of modern Britain and Bensalemite (Bacon's) new science—is better likened to David's empire than to the kingdom of Jesus. It seems, then, that however Christian it may be, Bensalem's destiny is unaffected by the difference between David and Jesus. From the standpoint of pious Christianity, which certainly characterizes the outward appearance of Bensalem, this difference is of paramount importance. We have already come to suspect that Bensalemite hospitality and humanity, presented in the guise of Christian charity, are something more than what first meets the eye. But the additional clues suggest that the problem of Bensalemite piety has something to do with Chris-

[24]*Bacon* wrote a history of the reign of Henry VII, BW, VI, 23–245. This work is important for understanding Bacon's understanding of the relation between science and the modern state. It will be interesting for the reader to compare the account of Henry's statesmanship with that of King David, whose kingdom was doomed by the prophecy of Nathan.

tianity itself: that the Bensalemites are *true* Christians, and that this faith and its twin, science, are infected with the same malady or problem. And one cannot but notice that the character who tells the narrator about Bensalem's Christian sexual morality is Joabin, namesake of the biblical Joab, who we know, of course, assisted David's sin against Uriah the Hittite, for which the rule of David's successors was cursed by Nathan's prophecy. And it is strange that the unit of linear measure in Bensalem is the "karan," a word whose Hebrew origin means "horn" and "ruler of the House of David."[25] All of this suggests that Bensalem, the true destiny of Christianity, is indifferent to the distinction between the old and the new covenants, which according to Christian doctrine consists (among other things) in the saving grace of Jesus, whose divine perfection contrasts with the sinful weakness of king David, the very first of the House of David. But before we can follow up these clues, which are developed only in the account of Bensalem's present, we must return to the account of the island's history.

To explain why Bensalem is unknown to the rest of the world, the priest relates first the story of the ancient Bensalemites' encounter with the old Atlantis, the one the priest calls "the great Atlantis," and then the story of the lawgiver Solamona, who brought science to what has become the new Atlantis. The first story explains why the world no longer comes to Bensalem, as it did "about three thousand years ago," and the second explains why the Bensalemites do not travel freely to the rest of the world, as they did before the reign of Solamona "about nineteen hundred years ago."[26]

Now according to the first story, the rest of the world once travelled often to Bensalem, for navigation was then "much greater than at this day." But the navigation to Bensalem from America ultimately disappeared, because of the divine revenge—in the form of a flood—that swept away the "proud enterprises" of the old Atlantians. And the navigation to Bensalem from other parts of the world

---

[25]*Ezekiel* 29:21; Psalms 132: 17. The verb form means to send out rays or display horns.
[26]*Below,* pp. 51–52, 56.

disappeared because of wars and the "natural revolution of time."[27] The proud enterprises of which the priest speaks refer to the conflict between the ancient Bensalemites under King Altabin and (the old) Atlantis, Peru, and Mexico, which were "mighty and proud kingdoms in arms, shipping, and riches" and which in the space of ten years attacked Europe and Bensalem. To be precise, it was Mexico that attacked Europe and Peru that attacked Bensalem. Plato, the "great man with you" mentioned by the priest, must therefore have been wrong to have reported in his account of Atlantis that the attackers of Europe were the old Atlantians, who must rather have minded their own business. And this is not all about which Plato was wrong, as closer inspection of the story reveals.

For contrary to Plato's account the priest says that the Athenians—if indeed they were even the ones who saved Europe—were no less ruthless than their attackers, as appears from the accounts of the fate of the invaders of Europe at the hands of the Athenians on the one hand and the bloodless victory and extraordinary clemency of Altabin on the other. And while in Plato's account it is unclear whether, after the great war, the world was destroyed by natural disaster or divine revenge (Zeus' righteous chastisement), the priest leaves no doubt at all. For what he first calls "divine revenge" he later calls quite simply a "main accident of time."[28] At any rate, in Plato's account earthquakes and floods soon after the Athenian victory destroyed the entire body of Athenian warriors and caused Atlantis to be swallowed by the sea. And the deeds did not live even in Athenian memory because Athens, along with the rest of the world except for Egypt, was subject to "many and diverse destructions of mankind, of which the greatest are by fire and water, and lesser by countless other means."[29] In Plato's account, Egyptian antiquity—and so the surviving memory of the Athenian deeds—depends on the Egyptians' piety, for although the Egyptians are immune from destructions

[27]*Below*, pp. 54–55.
[28]*Ibid.*
[29]*Timaeus* 21c1–23d1.

caused by rain and fire, because of the fortuitous presence of the Nile, they are not immune from earthquakes.[30] But according to the priest the Egyptians' antiquity was due simply to natural chance: he is careful to note that the "whole tract" referred to by Plato "is little subject to earthquakes.[31]

In other words, Plato was wrong to suggest the influence of divine revenge and Egyptian piety. Rather, between the time of Altabin and the time of King Solamona both Egypt and Bensalem survived by sheer good fortune the world's accidental destruction. After Solamona's reign, however, the Egyptians' luck ran out, for the Egyptians were among the peoples who once sailed to Bensalem but who have not done so since the ages after Altabin's time.[32] And after Solamona's reign, the Bensalemites used the powers of their natural science to withstand the natural revolutions of time. The Bensalemites did not at first need piety, because their king Altabin was too humane to have called forth divine revenge. So why did the island need Christianity? Perhaps to absolve the Bensalemites of guilt for (unmentioned) crimes committed before the reign of Altabin, or perhaps to absolve them of guilt for crimes committed in the pursuit of science: we know that Solamona's laws of secrecy may be far less humane than Altabin's deeds. But however deserved some divine revenge may have been, after the reign of Solamona, whose intention was "to give perpetuity to that which was in his time so happily established," the Bensalemites need not have feared any divine punishments of the kinds suffered by the rest of the world.[33] And we must not forget that the new science heralded by the Bensalemites does not blush at aiming to overcome mortality, the very wages of sin. Even more we are forced to wonder why the islands needed Christianity. Why, we ask, is science unable to provide its own moral justification?

[30]*Seth* Benardete, "On Plato's *Timaeus* and Timaeus' Science Fiction," *Interpretation*, 2 (Summer, 1971), p. 30; *Timaeus* 22c1–23a1, 25e8–d7.
[31]*Below*, pp. 53–54.
[32]*Below*, p. 52.
[33]*Below*, p. 56.

After hearing the stories of Altabin and Solamona the sailors declare themselves to be free men. No longer fearing the danger of their "utter perdition," they find the Bensalemites' hospitality toward strangers to be so free as to incline them to forget what was dear at home.[34] The nature of this freedom—both the strangers' and the Bensalemites'—is characterized in the speeches about Bensalem's present, of which the first describes Bensalem's ultimate promise and the second describes the regulation of that promise. The first speech is made by the narrator and describes the "Feast of the Family." This feast honors any Bensalemite man "that shall live to see thirty persons descended of his body alive together, and all above three years old."[35] The honor recognizes mere longevity and fecundity, not any excellence of soul. For while the state executes the orders of the father, who is called the Tirsan, to mend the vicious ways of any members of his family, there is no mention of any moral requirements to be met by the father himself. And yet according to the narrator the feast shows Bensalem to be "compounded of all goodness." In other words, Bensalem, described later as "the virgin of the world,"[36] is in fact dedicated to the preservation and generation of human bodies, and thus to the most intense pleasure that accompanies the procreation of human bodies. And the full meaning of this fact is disclosed by the second speech, made by the character named Joabin.

[34]*Below*, p. 6o. The sailors declare their liberation in the sixteenth paragraph of the story. The sixteenth Essay in the *Essays* treats atheism and is followed by the seventeenth Essay on superstition. The principle of the sixteenth Essay is that any belief about God is better than unbelief. The principle of the seventeenth Essay is that unbelief is better than wrong or heretical belief. Given that such a contradiction must always redound to the benefit of unbelief or atheism—no believer will *ever* pronounce the truth of atheism—Bacon in the *Essays* makes an argument for atheism. In the seventeenth paragraph the narrator forgets a difference between six and seven days. If Essays sixteen and seventeen come together as a true essay on atheism, the number of Essays is equal to the number of paragraphs in the *New Atlantis*.
[35]*Below*, p. 6o–64.
[36]*Below*, p. 66.

Prompted by having heard of the ceremony of the Tirsan, in which "nature did so much preside" and which concerns the propagation of families, the narrator asks Joabin about nuptial copulation in Bensalem: about the laws and customs concerning marriage and especially about whether the Bensalemites practice polygamy. Joabin's answer consists of a long speech about European and Bensalemite sexual morality and a brief mention of the "Adam and Eve's pools."[37]

According to Joabin, there could be no greater contrast than that between the chaste Bensalemites and the corrupt Europeans. Whereas Bensalem is the most chaste of nations, "free from all pollution and foulness" and "the virgin of the world," European sexual practices are debauched. The Europeans are so corrupt that they liken marriage to prostitution when they argue that marriage is "a remedy for unlawful concupiscence" and that prostitution, in which "sin is turned into art," is to be defended as preventive of greater evils. This, says Joabin, the Bensalemites consider a preposterous wisdom which they call "Lot's Offer." At this point the narrator interrupts to comment, as the widow of Sarepta did to Elias, that Joabin had come to remind the narrator of his sins. It is true, admits the narrator, that the Bensalemites are indeed more righteous than the Europeans. Joabin then continues by outlining the laws that govern marriage in Bensalem: there is no polygamy; there is a month wait between first interview and marriage or contract for marriage; marriage without parental consent is "mulcted in the inheritors" (i.e., the offspring of the marriage are penalized); and finally, "because of the many hidden defects in men and women's bodies," before a contract is made one friend of the potential bride and one friend of the potential groom are permitted to see the bride and groom "severally bathe naked" in the "Adam and Eve's pools" that are situated near every town.

Now Joabin's speech is the most important in the *New Atlantis*. For except for the scientist who judged the miracle, Joabin is the only post-Solamonic character de-

[37]*Below*, pp. 64–68.

scribed as wise. Not even the Father of Salomon's House
who makes the story's final speech is described as wise.
Moreover, not only is Joabin wise, but he is said to be
"learned . . . of great policy, and excellently seen in the
laws and customs of Bensalem." In other words, Joabin
is wise in matters of government and rule, and as such
he espouses his Jewish doctrines, and their compatibility
with Bensalemite Christianity, from considerations of
policy rather from any conviction of truth.[38] As a man
wise in general and wise in matters of policy, Joabin
transcends the myths of Bensalem. Of all the characters
in the story, only Joabin's wisdom is comprehensive, and
in this respect it certainly matches, if not surpasses, the
wisdom of the large-hearted founder and lawgiver Solamona. It is all the more important, then, that it is Joabin
whose speeches disclose the full nature of Bensalem's
promise to humanity.

   As Bacon has Joabin report it, the Bensalemites' lik-
ening of European opinions about sex and marriage to
the preposterous wisdom of "Lot's Offer" is subtly in-
appropriate. Joabin argues that the Europeans claim to
offer daughters (as prostitutes or as wives) to save other
daughters, the potential victims of unnatural lust and
rape. But Lot offered his virgin daughters to save male
strangers, an act of extraordinary hospitality.[39] Bacon has
Joabin express Bensalemite abhorrence of the offering
of daughters to save daughters, while pointing to and
remaining silent about the practice of offering daughters
for the sake of visiting strangers. This fact should certainly cause us to wonder about the priest's earlier remark that the sailors should feel free to make "any other
request" of the Bensalemites, which would not be re-
fused, and to note Bacon's silence about what is commonly thought to be the most pressing need felt by

[38]*The* narrator says of Joabin only that "surely this man . . . would
ever acknowledge" the basic reality of Jesus, not that he believed
in it, and of the "Jewish dreams" that must be set aside in order
to count Joabin wise the narrator says only that Joabin was desirous
"by tradition among the Jews there" to have them believed, not
that Joabin believed them himself. Below, pp. 64–66.
[39]*Genesis* 19:1–11.

sailors who are newly in port.[40] And when the narrator
says that Joabin's speech made him feel like the widow
of Sarepta we are reminded again of the harsh possibilities lurking behind Bensalem's outward hospitality.
For in fact the widow commented not just that Elias had
come to remind her of her sin, but also that she feared
him to have come to kill her son.[41]

These facts lead us to wonder about the extraordinary
institution of the Adam and Eve's pools. This institution
is intended to regulate nuptial copulation by preventing
the disappointments—and by implication the infideli-
ties—that might spring from the discovery after marriage
of the "many hidden defects in men and women's
bodies."[42] Joabin says that because it involves only in-
direct familiarity by way of friends, the Bensalemites'
version of naked prenuptial interview is superior to the
one devised by "one of your men," who included it in
an account of a feigned commonwealth. Now it is not
clear whether by his remark Joabin refers to Plato's *Laws*
or More's *Utopia*. But in either case it would seem that
the Bensalemites' practice is in fact the least workable.
For while in Plato's Magnesia boys and girls see each
other frequently and in common in naked play, and naked only within the limits set by a moderate sense of
shame, and while in Utopia the spouse is seen by the
prospective spouse, in Bensalem the prospective spouse
is viewed by a friend of the other partner.[43] What, we
are tempted to ask, can prevent a friend from making
a false report of bodily defect in order to secure the
intended partner for himself or herself, a result entirely
possible since Joabin does not specify the sex of the
reporting partner? And thus what is to prevent adulteries
inflamed by the desire of friends for the husbands and
wives of friends, desire grounded in familiar knowledge
of bodies?

According to Joabin the Bensalemites are restrained
from vice by religion and by "reverence of a man's self";

[40]*Below,* pp. 44–46.
[41]*I* Kings 17:18.
[42]*Below,* p. 68.
[43]*Plato, Laws* 771e; More, *Utopia,* "Their Marriage Customs."

that is, by conscience and self-love.[44] But Bensalemite
conscience is in fact not entirely scrupulous, as we know
from having examined their apparent charity and hospitality. Therefore, we cannot be confident that the Bensalemites' self-love is fully armored against mere self-
ishness. And just as these questions and doubts arise,
Joabin's remarks are interrupted so that we hear no more
about the matter, our attention being turned instead to
the disclosure of Salomon's House. The institution of the
pools presumes that the Bensalemites can be moved by
the desire for other, more beautiful bodies—why else
would they need to prevent fraud and disappointment?
But as presented the institution points more to problems
than to sweet marital bliss. It points to the licentious
powers of choosiness, the love of one's own, and the
desire for more, which reside at the very heart of Bensalemite "goodness."

The problems of Bensalem's erotic practice are high-
lighted by what is by now surely obvious: The only Bensalemite said to be wise in general and wise in policy,
and who reveals the astonishing practice of the pools,
is the namesake of the biblical Joab, David's strong lieu-
tenant in his rise to power and one of the most ruthless
men in the Old Testament. It was Joab who treacherously
assassinated Abner and Amassa, who murdered David's
son Absalom, and who carried out David's treachery
against Uriah the Hittite.[45] It was Joab, then, who assisted
in the murder of Uriah, and in the virtual theft of Uriah's
wife that resulted from David's glimpse of her bathing
naked. Thus it is almost a joke for Joabin to have been
the one to tell the narrator, and us, about the Bensalemite practice of having individuals see their friends'
prospective spouses "bathe naked."

But it is more than a joke. For with Joabin's speech
Bacon reminds us that scientific and Christian Bensalem
has not replaced the defects of King David with the
perfections of Jesus. Whatever the reasons for scientific
Bensalem's acceptance of Christianity, it is clear that
this Christianity is compatible with the immoral means

---

[44]*Below*, p. 68.
[45]*II* Samuel 2–1 Kings 2:35.

and problematic ends of science. In Bensalem neither science nor Christianity is a safeguard against excessive desire. In fact, in the absence of such a safeguard Bensalemite science might be said to base human endeavor on the very foundation of moral excess: the human body and the myriad desires that bring one body into conflict with others. From what we have learned so far, we might suggest the following summary of Bacon's ideas: The world needs science because the universal faith—Christianity—which has been the source of sectarian conflict, has been inadequate for human needs. And yet science is somehow a form of that universal faith because science attempts to attain that faith's central promise—the resurrection of the body along with the final salvation of the soul. But no such promise, however scientific, can obviate the problems of pride and excess, or be substituted for a teaching about harmony and moderation and the limits of human striving and ambition, in other words, a teaching about politics. In fact, such a promise is dangerous because it proffers the vain hope that in conquering death and scarcity we can escape the limits and problems of political life. Such a promise is blind to politics, and thus it can unwittingly encourage the most rambunctious political passions. And no renewal of science by a reformed or chastened Christianity can do any better, because a scientific faith would be no more attentive to the truths of political life. The world needs science because of the shortcomings of Christian faith. But science is no certain salvation for the world. Indeed, it is a risky venture because science requires the most beguiling, and potentially dangerous, form of belief: the belief that politics can be wholly replaced by wisdom. We are left to wonder, then, whether the wise man of the story, Joabin, is or is not like his namesake, and whether Bensalem does or does not overcome the weakness that called forth Nathan's prophecy of eventual doom.

The last part of the story presents the magnificent return of a Father of Salomon's House, his revelation to the narrator of the "true state" of the scientific establishment, and the Father's gift of a "great largess' or tip to the narrator and his fellows.[46] The Father's description of the scientific establishment reveals it to be dedicated to the study of the things above and below the earth and the things in between—that is, to the study of the visible whole. In particular, the fellows of the House of Salomon investigate all varieties of bodies that are useful for the preservation, longevity, and delight of human bodies, as well as the means for war and deceit. But they do not study the relations between human beings, who have both bodies *and* wills. In other words, despite the mundane orientation of Bensalemite science, it seems to have no place for the study of politics.

We might reasonably suspect that the absence of political science is merely apparent, for we know that according to Bacon it is a subject both secret and retired. And we know that in the House of Salomon the fellows take an "oath of secrecy" for the concealing of inventions and experiments that are not to be revealed to the state.[47] But political science is not so secret that its very existence cannot be mentioned: Bacon speaks explicitly about its secrecy in the *Advancement of Learning*. Moreover, the Father of Salomon's House does not hesitate to mention—and so reveal—the oath of secrecy about inventions and experiments. There is then nothing to have prevented the Father from listing political science among the other studies pursued in the House of Salomon. The fact is that political science is just not studied in the House of Salomon. And this is a surprise because the oath of secrecy concerns the concealing of inventions and experiments from the state, as if the House of Salomon were ultimately sovereign in Bensalem. Thus the scientists decide matters of secrecy, presuming to know enough about the state to know what can and cannot be revealed to it, but without having investigated politics *scientifically*, beyond the ken of common opinions and in a manner "secret and retired." In Bensalem—the model of scientific destiny—the natural scientists labor secretly in the shadow of a secrecy whose roots they do not grasp and whose master they do not recognize. The scientists'

[46]*Below*, pp. 68–83.
[47]*Below*, p. 82.

oath of secrecy points beyond the scientists' knowledge to someone learned in political science and thus able to rule with a clear understanding of nature and mankind.

It seems, then, that we are simply left to wonder about who rules wisely in Bensalem, as Rawley suggested in his opening note to the reader.[48] The scientists do not understand the ways of rule, but they vow (ignorantly) to keep the state in the dark about some inventions and experiments. Thus neither the scientific elite nor the political establishment can claim comprehensive wisdom about policy and government. But a likely candidate should be obvious to us, even if he has not been to the Bensalemites: the only man described as comprehensively wise—the inconspicuous Joabin. And from this fact we learn much of why Bensalem may be something less than "the best state or mould of a commonwealth." Rawley missed the important point: It is not that we learn nothing about modern scientific politics in the *New Atlantis*, but rather that we are forced to wonder why the scientists remain in the dark about the full range of possibilities illuminated by the new science, and why in Bensalem there will always be things of which the most ardent proponents of the new science will be unwilling to hear.

Despite Joabin's comprehensive and authoritative wisdom, he reveals the limits of the scientific project's capacity to overcome the "necessities and miseries of humanity." His appearance in the story symbolizes the moral ambiguity of Bensalemite science. The limits of the scientific project do not spring from the mortality of the human body. According to the new science, the body's corruptibility is the jealous act of a conquerable God, whose revenge we need no longer fear. Rather, the limits of the scientific project spring from the fact that human life is erotic, not just in the sense of the need and desire for reproduction and for long or even eternal life, but also in the constant search for the rare and the beautiful, the excellent and the preeminent, the daring and the great. The latter are desires that cause as much conflict as they do enrichment, and they cannot be satisfied by

[48]*Below*, p. 36.

the mere provision of material plenty and the indefinite prolongation of life. They cannot be managed by a politics based simply on the natural desires for self-preservation and commodious living and may as easily be inflamed, rather than soothed, by the promises of scientific conquest. In identifying political wisdom with the ambiguous character Joabin, Bacon tells us that while science opens the way for the conquest of nature on behalf of the many, it does not necessarily restrain the horizons of those few who are by nature conquerors. Modern science can promise openly to save mankind in a wholly new world. But the politics of science must be secret and retired because only the most resolute souls will be willing to embrace such a world with full knowledge of its moral risks and dangers.

But why does Bacon recommend to humanity a project that he cannot describe with complete frankness? Or perhaps more to the point, why does Bacon side with those who are willing to take great risks? Either Bacon was himself a lover of glory and dangerous conquests—more like Achilles than like Socrates—or else he thought that a charitable concern for the crisis of his times gave him no choice. If the latter is the case, then to understand it fully we would have to think carefully about Bacon's understanding of Christianity. And if the former, then we would have to investigate further his understanding of human virtue. In either case (and both may obtain, as well) his teaching provokes the questions faced by anyone who is both thoughtful and tempted by the charms of conquest: the questions concerning the nature of the gods, the essence of self-knowledge, the merits of action and contemplation, and, ultimately, the nature of and relations between wisdom, courage, justice, and moderation. But these are all encompassed by the more general question of the best regime and the ends of political life. About these questions Bacon gives us much food for thought in the *New Atlantis*. To this extent, then, it performs the most important function of any teaching about "the best state or mould of a commonwealth," which is to confront the reader with the most fundamental problems of private and public life. To follow up these problems—and thus to understand Bacon's place in the course of modernity—the reader must return to the whole of Bacon's corpus, and especially to his long, theoretical work *The Advancement of Learning*. But first the reader must work through the *New Atlantis* with great care. The present introduction aims to show the way to but not through Bacon's political teaching. Bacon intended the search for his teaching to produce the virtues it was designed to illuminate, and the reader must remember that it is debilitating and unseemly to stalk domesticated game.

note on the texts

The text of the *New Atlantis* is reprinted from the widely respected, standard edition of Bacon's works by Spedding, Ellis, and Heath. The original spelling and punctuation have everywhere been retained. The *New Atlantis* was first published by Rawley in 1627, after Bacon's death, and no original manuscript of the work remains. All the editions of the *New Atlantis* since 1627 have been reprints of Rawley's first edition, and fortunately there are no serious variations in the editions. The 1627 edition has "Solamona" and "Salomon's House," while the 1658 and 1670 editions, among others, have "Salomona" and "Solomon's House." Spedding follows the 1627 edition and is supported in this choice by the Latin translation (editions of 1638 and 1648). There is evidence that Bacon oversaw the Latin translation. Although Spedding is probably correct, the variation makes no difference at all in the meaning of the text. The 1627 and 1670 editions have "It so fell out, that there was in one of the boats, one of our wise men of the society of Salomon's House . . ." while the 1658 edition, which Spedding follows, has "It so fell out that there was in one of the boats one of the wise men of the society of Solomon's House . . ." (see pp. 47–48, below). In this instance, Spedding is perhaps wrong, since the 1627 and 1670 texts are supported by the Latin translation. But Spedding's text is not emended in this volume, because again, the variation makes no difference at all in the meaning of the text. The 1627 edition and Spedding have "came aboard" (see p. 38, below), while the 1658 edition has "made aboard"; and the 1627 edition has "that we knew he spake it . . ." (see p. 51, below), while the 1658 edition and Spedding have "that we knew that he spake it. . . ." These minor variations require no emendation. No critical problems in the text have so far been discovered that would warrant rejecting Spedding's text as the standard edition of the *New Atlantis*.

The Great Instauration was published originally in Latin in 1620 as the *Instauratio magna*. The translation used in this volume is Spedding's widely reprinted translation. Although Spedding's translation is considered to be the standard, it is sometimes loose, and two important corrections have been noted (see pp. 21, 32, below).

principal dates

THE
GREAT INSTAURATION

PROŒMIUM[1].

---

[1] *Preliminary* comment

FRANCIS OF VERULAM[2]

REASONED THUS WITH HIMSELF,

AND JUDGED IT TO BE FOR THE INTEREST OF THE PRESENT AND FUTURE GENERATIONS THAT THEY SHOULD BE MADE ACQUAINTED WITH HIS THOUGHTS.

Being convinced that the human intellect makes its own difficulties, not using the true helps which are at man's disposal soberly and judiciously; whence follows manifold ignorance of things, and by reason of that ignorance mischiefs innumerable; he[3] thought all trial should be made, whether that commerce[4] between the mind of man and the nature of things, which is more precious than anything on earth, or at least than anything that is of the earth, might by any means be restored to its perfect and original condition, or if that may not be, yet reduced[5] to a better condition than that in which it now is. Now that the errors which have hitherto prevailed, and which will prevail for ever, should (if the mind be left to go its own way), either by the natural force of the understanding or by help of the aids and instruments of Logic, one by one correct themselves, was a thing not to be hoped for: because the primary notions of things which the mind readily and passively imbibes, stores up, and accumulates (and it is from them that all the rest flow) are false, confused, and overhastily abstracted from the facts; nor are the secondary and subsequent notions less arbitrary and inconstant; whence it follows that the entire fabric of human reason which we employ in the inquisition of nature, is badly put together and built up, and like some magnificent struc-

[2]*Bacon* was Baron of Verulam.
[3]*Throughout* the Procemium Bacon refers to himself.
[4]*Intercourse*
[5]*Brought* back

1

ture without any foundation. For while men are occupied in admiring and applauding the false powers of the mind, they pass by and throw away those true powers, which, if it be supplied with the proper aids and can itself be content to wait upon nature instead of vainly affecting to overrule her, are within its reach. There was but one course left, therefore,—to try the whole thing anew upon a better plan, and to commence a total reconstruction of sciences, arts, and all human knowledge, raised upon the proper foundations. And this, though in the project and undertaking it may seem a thing infinite and beyond the powers of man, yet when it comes to be dealt with it will be found sound and sober, more so than what has been done hitherto. For of this there is some issue;[6] whereas in what is now done in the matter of science there is only a whirling round about, and perpetual agitation, ending where it began. And although he was well aware how solitary an enterprise it is, and how hard a thing to win faith and credit for, nevertheless he was resolved not to abandon either it or himself; nor to be deterred from trying and entering upon that one path which is alone open to the human mind. For better it is to make a beginning of that which may lead to something, than to engage in a perpetual struggle and pursuit in courses which have no exit. And certainly the two ways of contemplation are much like those two ways of action, so much celebrated, in this—that the one, arduous and difficult in the beginning, leads out at last into the open country; while the other, seeming at first sight easy and free from obstruction, leads to pathless and precipitous places.

Moreover, because he knew not how long it might be before these things would occur to any one else, judging especially from this, that he has found no man hitherto who has applied his mind to the like, he resolved to publish at once so much as he has been able to complete. The cause of which haste was not ambition for himself, but solicitude for the work; that in case of his death there might remain some outline and project of that which he had conceived, and some evidence likewise of

---

[6]*He* has already produced some results.

his honest mind and inclination towards the benefit of the human race. Certain it is that all other ambition whatsoever seemed poor in his eyes compared with the work which he had in hand; seeing that the matter at issue is either nothing, or a thing so great that it may well be content with its own merit, without seeking other recompence.

EPISTLE DEDICATORY.

JAMES,[7]

BY THE GRACE OF GOD
OF GREAT BRITAIN,
FRANCE, AND IRELAND
KING,

DEFENDER OF THE FAITH, ETC.

Most Gracious and Mighty King,

Your Majesty may perhaps accuse me of larceny, having stolen from your affairs so much time as was required for this work. I know not what to say for myself. For of time there can be no restitution, unless it be that what has been abstracted from your business may perhaps go to the memory of your name and the honour of your age; if these things are indeed worth anything. Certainly they are quite new; totally new in their very kind: and yet they are copied from a very ancient model; even the world itself and the nature of things and of the mind. And to say truth, I am wont for my own part to regard this work as a child of time rather than of wit;[8] the only wonder being that the first notion of the thing, and such great suspicions concerning matters long established, should have come into any man's mind. All the rest follows readily enough. And no doubt there is something of accident (as we call it) and luck as well in what men think as in what they do or say. But for this accident which I speak of, I wish that if there be any good in what I have to offer, it may be ascribed to the infinite

[7]*James* I, King of England, 1603–1625
[8]A product of favorable circumstances rather than of Bacon's intellect

5

mercy and goodness of God, and to the felicity of your Majesty's times; to which as I have been an honest and affectionate servant in my life, so after my death I may yet perhaps, through the kindling of this new light in the darkness of philosophy, be the means of making this age famous to posterity; and surely to the times of the wisest and most learned of kings belongs of right the regeneration and restoration of the sciences. Lastly, I have a request to make—a request no way unworthy of your Majesty, and which especially concerns the work in hand; namely, that you who resemble Solomon in so many things—in the gravity of your judgments, in the peacefulness of your reign, in the largeness of your heart, in the noble variety of the books which you have composed—would further follow his example[9] in taking order[10] for the collecting and perfecting of a Natural and Experimental History, true and severe[11] (unincumbered with literature and book-learning), such as philosophy may be built upon,—such, in fact, as I shall in its proper place describe: that so at length, after the lapse of so many ages, philosophy and the sciences may no longer float in air, but rest on the solid foundation of experience of every kind, and the same well examined and weighed. I have provided the machine, but the stuff must be gathered from the facts of nature. May God Almighty long preserve your Majesty!

> Your Majesty's
> > Most bounden and devoted
> > Servant,
>
> **FRANCIS VERULAM,**
> **Chancellor**

[9]**See** I Kings 4:33 (Heb. 5:13).
[10]**To** command that it be done
[11]**Serious,** accurate

THE GREAT INSTAURATION.[12]

PREFACE.

That the state of knowledge is not prosperous nor greatly advancing; and that a way must be opened for the human understanding entirely different from any hitherto known, and other helps provided, in order that the mind may exercise over the nature of things the authority which properly belongs to it.

It seems to me that men do not rightly understand either their store[13] or their strength, but overrate the one and underrate the other. Hence it follows, that either from an extravagant estimate of the value of the arts which they possess, they seek no further; or else from too mean an estimate of their own powers, they spend their strength in small matters and never put it fairly to the trial in those which go to the main.[14] These are as the pillars of fate set in the path of knowledge;[15] for men have neither desire nor hope to encourage them to penetrate further. And since opinion of store[16] is one of the chief causes of want, and satisfaction with the present induces neglect of provision for the future, it becomes a thing not only useful, but absolutely necessary, that the excess of honour and admiration with which our existing stock of inventions is regarded be in the very entrance and threshold of the work, and that frankly

[12]***Restoration*** after decay; setting up or founding
[13]***Tools***
[14]***The*** most important matters
[15]***These*** defects act as the limits of human effort.
[16]***Plenty***

and without circumlocution, stripped off, and men be duly warned not to exaggerate or make too much of them. For let a man look carefully into all that variety of books with which the arts and sciences abound, he will find everywhere endless repetitions of the same thing, varying in the method of treatment, but not new in substance, insomuch that the whole stock, numerous as it appears at first view, proves on examination to be but scanty. And for its value and utility it must be plainly avowed that that wisdom which we have derived principally from the Greeks is but like the boyhood of knowledge, and has the characteristic property of boys: it can talk, but it cannot generate; for it is fruitful of controversies but barren of works. So that the state of learning as it now is appears to be represented to the life in the old fable of Scylla, who had the head and face of a virgin, but her womb was hung round with barking monsters, from which she could not be delivered. For in like manner the sciences to which we are accustomed have certain general positions which are specious and flattering; but as soon as they come to particulars, which are as the parts of generation, when they should produce fruit and works, then arise contentions and barking disputations, which are the end of the matter and all the issue they can yield. Observe also, that if sciences of this kind had any life in them, that could never have come to pass which has been the case now for many ages—that they stand almost at a stay,[17] without receiving any augmentations worthy of the human race; insomuch that many times not only what was asserted once is asserted still, but what was a question once is a question still, and instead of being resolved by discussion is only fixed and fed; and all the tradition and succession of schools is still a succession of masters and scholars, not of inventors and those who bring to further perfection the things invented. In the mechanical arts we do not find it so; they, on the contrary, as having in them some breath of life, are continually growing and becoming more perfect. As originally invented they are commonly rude, clumsy, and shapeless; afterwards they acquire new pow-

[17]*Stop*

ers and more commodious arrangements and construc-
tions; in so far that men shall sooner leave the study
and pursuit of them and turn to something else, than
they arrive at the ultimate perfection of which they are
capable.[18] Philosophy and the intellectual sciences, on
the contrary, stand like statues, worshipped and celebrated, but not moved or advanced. Nay, they sometimes
flourish most in the hands of the first author, and afterwards degenerate. For when men have once made
over their judgments to others' keeping, and (like those
senators whom they called *Pedarii*)[19] have agreed to sup-
port some one person's opinion, from that time they
make no enlargement of the sciences themselves, but
fall to the servile office of embellishing certain individual
authors and increasing their retinue. And let it not be
said that the sciences have been growing gradually till
they have at last reached their full stature, and so (their
course being completed) have settled in the works of a
few writers; and that there being now no room for the
invention of better, all that remains is to embellish and
cultivate those things which have been invented already.
Would it were so! But the truth is that this appropriating[20]
of the sciences has its origin in nothing better than the
confidence of a few persons and the sloth and indolence
of the rest. For after the sciences had been in several
parts perhaps cultivated and handled diligently, there
has risen up some man of bold disposition, and famous
for methods and short ways which people like, who has
in appearance reduced them to an art, while he has in
fact only spoiled all that the others had done. And yet
this is what posterity like, because it makes the work
short and easy, and saves further inquiry, of which they
are weary and impatient. And if any one take this general
acquiescence and consent for an argument of weight, as
being the judgment of Time, let me tell him that the
reasoning on which he relies is most fallacious and weak.
For, first, we are far from knowing all that in the matter

[18]***Bacon*** means that the mechanical arts are continually progres-
sive; they do not quickly come to a specious perfection.
[19]***Roman*** senators of inferior rank
[20]***Making*** over to a special owner

of sciences and arts has in various ages and places been brought to light and published; much less, all that has been by private persons secretly attempted and stirred; so neither the births nor the miscarriages of Time are entered in our records. Nor, secondly, is the consent itself and the time it has continued a consideration of much worth. For however various are the forms of civil polities, there is but one form of polity in the sciences; and that always has been and always will be popular. Now the doctrines which find most favour with the populace are those which are either contentious and pugnacious, or specious and empty; such, I say, as either entangle[21] assent or tickle it. And therefore no doubt the greatest wits in each successive age have been forced out of their own course; men of capacity and intellect above the vulgar having been fain,[22] for reputation's sake, to bow to the judgment of the time and the multitude; and thus if any contemplations of a higher order took light anywhere, they were presently blown out by the winds of vulgar opinions. So that Time is like a river, which has brought down to us things light and puffed up, while those which are weighty and solid have sunk. Nay, those very authors who have usurped a kind of dictatorship in the sciences and taken upon them to lay down the law with such confidence, yet when from time to time they come to themselves again, they fall to complaints of the subtlety of nature, the hiding-places of truth, the obscurity of things, the entanglement of causes, the weakness of the human mind; wherein nevertheless they show themselves never the more modest, seeing that they will rather lay the blame upon the common condition of men and nature than upon themselves. And then whatever any art fails to attain, they ever set it down upon the authority of that art itself as impossible of attainment; and how can art be found guilty when it is judge in its own cause? So it is but a device for exempting ignorance from ignominy. Now for those things which are delivered and received, this is their condition: barren of works, full of questions; in point of enlarge-

---

[21]*Catch*, ensnare
[22]*Willing*, inclined, happy

ment slow and languid; carrying a show of perfection in the whole, but in the parts ill filled up; in selection popular, and unsatisfactory even to those who propound them; and therefore fenced round and set forth with sundry artifices. And if there be any who have determined to make trial for themselves, and put their own strength to the work of advancing the boundaries of the sciences, yet have they not ventured to cast themselves completely loose from received opinions or to seek their knowledge at the fountain;[23] but they think they have done some great thing if they do but add and introduce into the existing sum of science something of their own; prudently considering with themselves that by making the addition they can assert their liberty, while they retain the credit of modesty by assenting to the rest. But these mediocrities and middle ways so much praised, in deferring to opinions and customs, turn to the great detriment of the sciences. For it is hardly possible at once to admire an author and to go beyond him; knowledge being as water, which will not rise above the level from which it fell. Men of this kind, therefore, amend some things, but advance little; and improve the condition of knowledge, but do not extend its range. Some, indeed, there have been who have gone more boldly to work, and taking it all for an open matter and giving their genius full play, have made a passage for themselves and their own opinions by pulling down and demolishing former ones; and yet all their stir has but little advanced the matter; since their aim has been not to extend philosophy and the arts in substance and value, but only to change doctrines and transfer the kingdom of opinions to themselves; whereby little has indeed been gained, for though the error be the opposite of the other, the causes of erring are the same in both. And if there have been any who, not binding themselves either to other men's opinions or to their own, but loving liberty, have desired to engage others along with themselves in search, these, though honest in intention, have been weak in endeavour. For they have been content to follow

[23]*The* source

probable[24] reasons, and are carried round in a whirl of arguments, and in the promiscuous liberty of search have relaxed the severity of inquiry. There is none who has dwelt upon experience and the facts of nature as long as is necessary. Some there are indeed who have committed themselves to the waves of experience, and almost turned mechanics;[25] yet these again have in their very experiments pursued a kind of wandering inquiry, without any regular system of operations. And besides they have mostly proposed to themselves certain petty tasks, taking it for a great matter to work out some single discovery;—a course of proceeding at once poor in aim and unskilful in design. For no man can rightly and successfully investigate the nature of anything in the thing itself; let him vary his experiments as laboriously as he will, he never comes to a resting-place, but still finds something to seek beyond.[26] And there is another thing to be remembered; namely, that all industry in experimenting has begun with proposing to itself certain definite works to be accomplished, and has pursued them with premature and unseasonable eagerness; it has sought, I say, experiments of Fruit, not experiments of Light; not imitating the divine procedure, which in its first day's work created light only and assigned to it one entire day; on which day it produced no material work, but proceeded to that on the days following. As for those who have given the first place to Logic, supposing that the surest helps to the sciences were to be found in that, they have indeed most truly and excellently perceived that the human intellect left to its own course is not to be trusted; but then the remedy is altogether too weak for the disease; nor is it without evil in itself. For the Logic which is received, though it be very properly applied to civil business and to those arts which rest in discourse and opinion, is not nearly subtle enough to deal with nature; and in offering at[27] what it cannot

[24]*Only* apparently true, only likely
[25]*Practitioners* or students of the practical arts and sciences
[26]*In* this sentence Bacon means that the phenomena of nature are interconnected.
[27]*Aiming* at

master, has done more to establish and perpetuate error than to open the way to truth.

Upon the whole therefore, it seems that men have not been happy hitherto either in the trust which they have placed in others or in their own industry with regard to the sciences; especially as neither the demonstrations nor the experiments as yet known are much to be relied upon. But the universe to the eye of the human understanding is framed like a labyrinth; presenting as it does on every side so many ambiguities of way, such deceitful resemblances of objects and signs, natures so irregular in their lines, and so knotted and entangled. And then the way is still to be made by the uncertain light of the sense, sometimes shining out, sometimes clouded over, through the woods of experience and particulars; while those who offer themselves for guides are (as was said) themselves also puzzled, and increase the number of errors and wanderers. In circumstances so difficult neither the natural force of man's judgment nor even any accidental felicity[28] offers any chance of success. No excellence of wit, no repetition of chance experiments, can overcome such difficulties as these. Our steps must be guided by a clue, and the whole way from the very first perception of the senses must be laid out upon a sure plan. Not that I would be understood to mean that nothing whatever has been done in so many ages by so great labours. We have no reason to be ashamed of the discoveries which have been made, and no doubt the ancients proved themselves in everything that turns on wit and abstract meditation, wonderful men. But as in former ages when men sailed only by observation of the stars, they could indeed coast along the shores of the old continent or cross a few small and mediterranean[29] seas; but before the ocean could be traversed and the new world discovered, the use of the mariner's needle, as a more faithful and certain guide, had to be found out; in like manner the discoveries which have been hitherto made in the arts and sciences are such as might be made by practice, meditation, obser-

[28]*Accidental* good result
[29]*Inland*

vation, argumentation,—for they lay near to the senses, and immediately beneath common notions; but before we can reach the remoter and more hidden parts of nature, it is necessary that a more perfect use and application of the human mind and intellect be introduced.

For my own part at least, in obedience to the everlasting love of truth, I have committed myself to the uncertainties and difficulties and solitudes of the ways, and relying on the divine assistance have upheld my mind both against the shocks and embattled ranks of opinion, and against my own private and inward hesitations and scruples, and against the fogs and clouds of nature, and the phantoms flitting about on every side; in the hope of providing at last for the present and future generations guidance more faithful and secure. Wherein if I have made any progress, the way has been opened to me by no other means than the true and legitimate humiliation of the human spirit. For all those who before me have applied themselves to the invention of arts have but cast a glance or two upon facts and examples and experience, and straightway proceeded, as if invention were nothing more than an exercise of thought, to invoke their own spirits to give them oracles. I, on the contrary, dwelling purely and constantly among the facts of nature, withdraw my intellect from them no further than may suffice to let the images and rays of natural objects meet in a point, as they do in the sense of vision; whence it follows that the strength and excellency of the wit has but little to do in the matter. And the same humility which I use in inventing I employ likewise in teaching. For I do not endeavour either by triumphs of confutation, or pleadings of antiquity, or assumption of authority, or even by the veil of obscurity, to invest these inventions of mine with any majesty; which might easily be done by one who sought to give lustre to his own name rather than light to other men's minds. I have not sought (I say) nor do I seek either to force or ensnare men's judgments, but I lead them to things themselves and the concordances[30] of things, that they may see for themselves what they have, what they can dispute, what they

[30]*Agreements,* harmonies, connections

can add and contribute to the common stock. And for myself, if in anything I have been either too credulous or too little awake and attentive, or if I have fallen off by the way and left the inquiry incomplete, nevertheless I so present these things naked and open, that my errors can be marked and set aside before the mass of knowledge be further infected by them; and it will be easy also for others to continue and carry on my labours. And by these means I suppose that I have established for ever a true and lawful marriage between the empirical and the rational faculty, the unkind and ill-starred divorce and separation of which has thrown into confusion all the affairs of the human family.

Wherefore, seeing that these things do not depend upon myself, at the outset of the work I most humbly and fervently pray to God the Father, God the Son, and God the Holy Ghost, that remembering the sorrows of mankind and the pilgrimage of this our life wherein we wear out days few and evil, they will vouchsafe through my hands to endow the human family with new mercies. This likewise I humbly pray, that things human may not interfere with things divine, and that from the opening of the ways of sense[31] and the increase of natural light there may arise in our minds no incredulity or darkness with regard to the divine mysteries; but rather that the understanding being thereby purified and purged of fancies and vanity, and yet not the less subject and entirely submissive to the divine oracles, may give to faith that which is faith's. Lastly, that knowledge being now discharged of that venom which the serpent infused into it, and which makes the mind of man to swell, we may not be wise above measure and sobriety, but cultivate truth in charity.

And now having said my prayers I turn to men; to whom I have certain salutary admonitions to offer and certain fair requests to make. My first admonition (which was also my prayer) is that men confine the sense within the limits of duty in respect of things divine: for the sense is like the sun, which reveals the face of earth, but seals and shuts up the face of heaven. My next, that

[31]*The* renewed investigation of empirical evidence

in flying from this evil they fall not into the opposite error, which they will surely do if they think that the inquisition of nature is in any part interdicted or forbidden. For it was not that pure and uncorrupted natural knowledge whereby Adam gave names to the creatures according to their propriety,[32] which gave occasion to the fall. It was the ambitious and proud desire of moral knowledge to judge of good and evil, to the end that man may revolt from God and give laws to himself, which was the form and manner of the temptation. Whereas of the sciences which regard nature, the divine philosopher[33] declares that "it is the glory of God to conceal a thing, but it is the glory of the King to find a thing out."[34] Even as though the divine nature took pleasure in the innocent and kindly sport of children playing at hide and seek, and vouchsafed of his kindness and goodness to admit the human spirit for his playfellow at that game. Lastly, I would address one general admonition to all; that they consider what are the true ends of knowledge, and that they seek it not either for pleasure of the mind, or for contention, or for superiority to others, or for profit, or fame, or power, or any of these inferior things; but for the benefit and use of life; and that they perfect and govern it in charity. For it was from lust of power that the angels fell, from lust of knowledge that man fell; but of charity there can be no excess, neither did angel or man ever come in danger by it.

The requests I have to make are these. Of myself I say nothing; but in behalf of the business which is in hand I entreat men to believe that it is not an opinion to be held, but a work to be done; and to be well assured that I am labouring to lay the foundation, not of any sect or doctrine, but of human utility and power. Next, I ask them to deal fairly by their own interests, and laying aside all emulations[35] and prejudices in favor of this or that opinion, to join in consultation for the com-

[32]*Particular* nature or character
[33]*Solomon*
[34]*Proverbs* 25:2.
[35]*Ambitious* rivalries for power

mon good; and being now freed and guarded by the securities and helps which I offer from the errors and impediments of the way, to come forward themselves and take part in that which remains to be done. Moreover, to be of good hope, nor to imagine that this Instauration of mine is a thing infinite and beyond the power of man, when it is in fact the true end and termination of infinite error; and seeing also that it is by no means forgetful of the conditions of mortality and humanity, (for it does not suppose that the work can be altogether completed within one generation, but provides for its being taken up by another); and finally that it seeks for the sciences not arrogantly in the little cells of human wit, but with reverence in the greater world. But it is the empty things that are vast: things solid are most contracted and lie in little room.[36] And now I have only one favour more to ask (else injustice to me may perhaps imperil the business itself)—that men will consider well how far, upon that which I must needs assert (if I am to be consistent with myself), they are entitled to judge and decide upon these doctrines of mine; inasmuch as all that premature human reasoning which anticipates inquiry, and is abstracted from the facts rashly and sooner than is fit, is by me rejected (so far as the inquisition of nature is concerned), as a thing uncertain, confused, and ill built up; and I cannot be fairly asked to abide by the decision of a tribunal which is itself on its trial.

[36]*The* task of the Instauration is possible

The Plan of the Work.

The Work Is In Six Parts:—

1. *The Divisions of the Sciences.*
2. *The New Organon[37]; or Directions concerning the Interpretation of Nature.*
3. *The Phenomena of the Universe; or a Natural and Experimental History for the foundation of Philosophy.*
4. *The Ladder of the Intellect.*
5. *The Forerunners; or Anticipations of the New Philosophy.*
6. *The New Philosophy; or Active Science.*

The Arguments of the several Parts.

It being part of my design to set everything forth, as far as may be, plainly and perspicuously (for nakedness of the mind is still, as nakedness of the body once was, the companion of innocence and simplicity), let me first explain the order and plan of the work. I distribute it into six parts.

The first part[38] exhibits a summary or general description of the knowledge which the human race at present possesses. For I thought it good to make some pause upon that which is received; that thereby the old may be more easily made perfect and the new more easily approached. And I hold the improvement of that which we have to be as much an object as the acquisition of more. Besides which it will make me the better listened to; for "He that is ignorant (says the proverb)

[37]*An* instrument of thought or knowledge; also the traditional title of Aristotle's logical treatises

[38]*Part* one is elsewhere declared to be wanting although "some account [of the divisions of the sciences] will be found in the 'Second Book of the Proficience and Advancement of Learning' "; below, p. 33.

receives not the words of knowledge, unless thou first tell him that which is in his own heart." We will therefore make a coasting voyage along the shores of the arts and sciences received; not without importing into them some useful things by the way.

In laying out the divisions of the sciences however, I take into account not only things already invented and known, but likewise things omitted which ought to be there. For there are found in the intellectual as in the terrestial globe waste regions as well as cultivated ones. It is no wonder therefore if I am sometimes obliged to depart from the ordinary divisions. For in adding to the total you necessarily alter the parts and sections; and the received divisions of the sciences are fitted only to the received sum of them as it stands now.

With regard to those things which I shall mark as omitted, I intend not merely to set down a simple title or a concise argument of that which is wanted. For as often as I have occasion to report anything as deficient, the nature of which is at all obscure, so that men may not perhaps easily understand what I mean or what the work is which I have in my head, I shall always (provided it be a matter of any worth) take care to subjoin either directions for the execution of such work, or else a portion of the work itself executed by myself as a sample of the whole: thus giving assistance in every case either by work or by counsel. For if it were for the sake of my own reputation only and other men's interests were not concerned in it, I would not have any man think that in such cases merely some light and vague notion has crossed my mind, and that the things which I desire and offer at are no better than wishes; when they are in fact things which men may certainly command if they will, and of which I have formed in my own mind a clear and detailed conception. For I do not propose merely to survey these regions in my mind, like an augur taking auspices, but to enter them like a general who means to take possession.—So much for the first part of the work.

Having thus coasted past the ancient arts, the next point is to equip the intellect for passing beyond. To the second part[39] therefore belongs the doctrine concerning the better and more perfect use of human reason in the inquisition of things, and the true helps of the understanding: that thereby (as far as the condition of mortality and humanity allows) the intellect may be raised and exalted, and made capable of overcoming the difficulties and obscurities of nature. The art which I introduce with this view (which I call *Interpretation of Nature*) is a kind of logic; though the difference between it and the ordinary logic is great; indeed immense. For the ordinary logic professes to contrive and prepare helps and guards for the understanding, as mine does; and in this one point they agree. But mine differs from it in three points especially; viz. in the end aimed at; in the order of demonstration; and in the starting point of the inquiry.

For the end which this science of mine proposes is the invention not of arguments but of arts; not of things in accordance with principles, but of principles themselves; not of probable reasons, but of designations and directions for works. And as the intention is different, so accordingly is the effect; the effect of the one being to overcome an opponent in argument, of the other to command[40] nature in action.

In accordance with this end is also the nature and order of the demonstrations. For in the ordinary logic almost all the work is spent about the syllogism.[41] Of induction[42] the logicians seem hardly to have taken any serious thought, but they pass it by with a slight notice, and hasten on to the formulae of disputation. I on the contrary reject demonstration by syllogism, as acting too confusedly, and letting nature slip out of its hands. For although no one can doubt that things which agree in a middle term[43] agree with one another (which is a prop-

---

[39]*The* second part is provided in the *Novum Organum,* but only as "a summary digested into aphorisms"; ibid.

[40]*The* Latin is *vincitur,* which should be translated here as "conquer."

[41]*The* scheme of deductive logic composed of a major premise (every animal is mortal), a minor premise (every man is an animal), and a conclusion (every man is mortal)

[42]*Reasoning* from particulars to general principles

[43]*The* term common to the two premises of a syllogism, *i.e.,* "animal" in the example in n. 41, above

osition of mathematical certainty), yet it leaves an opening for deception; which is this. The syllogism consists
of propositions; propositions of words; and words are the
tokens and signs of notions. Now if the very notions of
the mind (which are as the soul of words and the basis
of the whole structure) be improperly and overhastily
abstracted from facts, vague, not sufficiently definite,
faulty in short in many ways, the whole edifice tumbles.
I therefore reject the syllogism; and that not only as
regards principles (for to principles the logicians themselves do not apply it) but also as regards middle propositions;[44] which, though obtainable no doubt by the
syllogism, are, when so obtained, barren of works, remote from practice, and altogether unavailable for the
active department of the sciences. Although therefore
I leave to the syllogism and these famous and boasted
modes of demonstration their jurisdiction over popular
arts and such as are matter of opinion (in which department I leave all as it is), yet in dealing with the
nature of things I use induction throughout, and that in
the minor propositions[45] as well as the major. For I consider induction to be that form of demonstration which
upholds the sense, and closes[46] with nature, and comes
to the very brink of operation, if it does not actually
deal with it.

Hence it follows that the order of demonstration is
likewise inverted. For hitherto the proceeding has been
to fly at once from the sense[47] and particulars up to the
most general propositions, as certain fixed poles for the
argument to turn upon, and from these to derive the
rest by middle terms: a short way, no doubt, but pre-
cipitate; and one which will never lead to nature, though
it offers an easy and ready way to disputation. Now my
plan is to proceed regularly and gradually from one ax-
iom to another, so that the most general are not reached

[44]**Propositions** about the middle term of a syllogism. In a syllogism,
the middle term disappears from the conclusion.
[45]**Premises.** Bacon means here that he will apply inductive reasoning to the propositions of major and minor premises.
[46]**Comes** into contact with
[47]**Empirical** evidence

till the last: but then when you do come to them you find them to be not empty notions, but well defined, and such as nature would really recognise as her first principles, and such as lie at the heart and marrow of things.

But the greatest change I introduce is in the form itself of induction and the judgment made thereby. For the induction of which the logicians speak, which proceeds by simple enumeration, is a puerile thing; concludes at hazard; is always liable to be upset by a contradictory instance; takes into account only what is known and ordinary; and leads to no result.

Now what the sciences stand in need of is a form of induction which shall analyse experience and take it to pieces, and by a due process of exclusion and rejection lead to an inevitable conclusion. And if that ordinary mode of judgment practised by the logicians was so laborious, and found exercise for such great wits, how much more labour must we be prepared to bestow upon this other, which is extracted not merely out of the depths of the mind, but out of the very bowels of nature.

Nor is this all. For I also sink the foundations of the sciences deeper and firmer; and I begin the inquiry nearer the source than men have done heretofore; submitting to examination those things which the common logic takes on trust. For first, the logicians borrow the principles of each science from the science itself; secondly, they hold in reverence the first notions of the mind; and lastly, they receive as conclusive the immediate informations of the sense, when well disposed. Now upon the first point, I hold that true logic ought to enter the several provinces of science armed with a higher authority than belongs to the principles of those sciences themselves, and ought to call those putative principles to account until they are fully established. Then with regard to the first notions of the intellect; there is not one of the impressions taken by the intellect when left to go its own way, but I hold it for suspected, and no way established, until it has submitted to a new trial and a fresh judgment has been thereupon pronounced. And lastly, the information of the sense itself I sift and examine in many ways. For certain it is that the senses

deceive; but then at the same time they supply the means
of discovering their own errors; only the errors are here,
the means of discovery are to seek.

The sense fails in two ways. Sometimes it gives no
information, sometimes it gives false information. For
first, there are very many things which escape the sense,
even when best disposed and no way obstructed; by
reason either of the subtlety of the whole body, or the
minuteness of the parts, or distance of place, or slowness
or else swiftness of motion, or familiarity of the object,
or other causes. And again when the sense does appre-
hend a thing its apprehension is not much to be relied
upon. For the testimony and information of the sense
has reference always to man, not to the universe; and
it is a great error to assert that the sense is the measure
of things.

To meet these difficulties, I have sought on all sides
diligently and faithfully to provide helps for the sense—
substitutes to supply its failures, rectifications to correct
its errors; and this I endeavour to accomplish not so
much by instruments as by experiments. For the subtlety
of experiments is far greater than that of the sense itself,
even when assisted by exquisite instruments; such experiments, I mean, as are skilfully and artificially devised
for the express purpose of determining the point in question. To the immediate and proper perception of the
sense therefore I do not give much weight; but I contrive
that the office[48] of the sense shall be only to judge of
the experiment, and that the experiment itself shall judge
of the thing. And thus I conceive that I perform the
office of a true priest of the sense (from which all knowledge in nature must be sought, unless men mean to go
mad) and a not unskilful interpreter of its oracles; and
that while others only profess to uphold and cultivate
the sense, I do so in fact. Such then are the provisions
I make for finding the genuine light of nature and kindling and bringing it to bear. And they would be sufficient of themselves, if the human intellect were even,
and like a fair sheet of paper with no writing on it. But
since the minds of men are strangely possessed and beset,

[48]*Function*

so that there is no true and even surface left to reflect the genuine rays of things, it is necessary to seek a remedy for this also.

Now the idols, or phantoms, by which the mind is occupied are either adventitious or innate. The adventitious come into the mind from without; namely, either from the doctrines and sects of philosophers, or from perverse rules of demonstration. But the innate are inherent in the very nature of the intellect, which is far more prone to error than the sense is. For let men please themselves as they will in admiring and almost adoring the human mind, this is certain: that as an uneven mirror distorts the rays of objects according to its own figure and section, so the mind, when it receives impressions of objects through the sense, cannot be trusted to report them truly, but in forming its notions mixes up its own nature with the nature of things.

And as the first two kinds of idols are hard to eradicate, so idols of this last kind cannot be eradicated at all. All that can be done is to point them out, so that this insidious action of the mind may be marked and reproved (else as fast as old errors are destroyed new ones will spring up out of the ill complexion[49] of the mind itself, and so we shall have but a change of errors, and not a clearance); and to lay it down once for all as a fixed and established maxim, that the intellect is not qualified to judge except by means of induction, and induction in its legitimate form. This doctrine then of the expurgation of the intellect to qualify it for dealing with truth, is comprised in three refutations: the refutation of the Philosophies; the refutation of the Demonstrations; and the refutation of the Natural Human Reason.[50] The explanation of which things, and of the true relation between the nature of things and the nature of the mind, is as

---

[49]*Nature,* disposition

[50]*Bacon's* considerations of these "refutations" are scattered throughout his writings. See Anderson, *The Philosophy of Francis Bacon,* pp. 40–47. For the "refutation of the Philosophies," see Farrington, *The Philosophy of Francis Bacon,* which contains translations of *The Masculine Birth of Time, Thoughts and Conclusions,* and *The Refutation of Philosophies.*

the strewing and decoration of the bridal chamber of the Mind and the Universe, the Divine Goodness assisting; out of which marriage let us hope (and be this the prayer of the bridal song) there may spring helps to man, and a line and race of inventions that may in some degree subdue and overcome the necessities and miseries of humanity. This is the second part of the work.

But I design not only to indicate and mark out the ways, but also to enter them. And therefore the third part[51] of the work embraces the Phenomena of the Universe; that is to say, experience of every kind, and such a natural history as may serve for a foundation to build philosophy upon. For a good method of demonstration or form of interpreting nature may keep the mind from going astray or stumbling, but it is not any excellence of method that can supply it with the material of knowledge. Those however who aspire not to guess and divine, but to discover and know; who propose not to devise mimic[52] and fabulous worlds of their own, but to examine and dissect the nature of this very world itself; must go to facts themselves for everything. Nor can the place of this labour and search and worldwide perambulation be supplied by any genius or meditation or argumentation; no, not if all men's wits could meet in one. This therefore we must have, or the business must be for ever abandoned. But up to this day such has been the condition of men in this matter, that it is no wonder if nature will not give herself into their hands.

For first, the information of the sense itself, sometimes failing, sometimes false; observation; careless, irregular, and led by chance; tradition, vain and fed on rumour; practice, slavishly bent upon its work; experiment, blind, stupid, vague, and prematurely broken off; lastly, natural history trivial and poor;—all these have contributed to supply the understanding with very bad materials for philosophy and the sciences.

[51]*Although* incomplete, the third part is the most filled out. See Anderson, *The Philosophy of Francis Bacon*, pp. 34–36, 40.
[52]*Imitation*

Then an attempt is made to mend the matter by a preposterous subtlety and winnowing of argument. But this comes too late, the case being already past remedy; and is far from setting the business right or sifting away the errors. The only hope therefore of any greater increase or progress lies in a reconstruction of the sciences.

Of this reconstruction the foundation must be laid in natural history, and that of a new kind and gathered on a new principle. For it is in vain that you polish the mirror if there are no images to be reflected; and it is as necessary that the intellect should be supplied with fit matter to work upon, as with safeguards to guide its working. But my history differs from that in use (as my logic does) in many things,—in end and office, in mass and composition, in subtlety, in selection also and setting forth, with a view to the operations which are to follow.

For first, the object of the natural history which I propose is not so much to delight with variety of matter or to help with present use of experiments, as to give light to the discovery of causes and supply a suckling philosophy with its first food. For though it be true that I am principally in pursuit of works and the active department of the sciences, yet I wait for harvest-time, and do not attempt to mow the moss[53] or to reap the green corn. For I well know that axioms once rightly discovered will carry whole troops of works along with them, and produce them, not here and there one, but in clusters. And that unseasonable and puerile hurry to snatch by way of earnest at the first works which come within reach, I utterly condemn and reject, as an Atalanta's apple[54] that hinders the race. Such then is the office of this natural history of mine.

Next, with regard to the mass and composition of it: I mean it to be a history not only of nature free and at large (when she is left to her own course and does her work her own way),—such as that of the heavenly bodies,

[53]*Cotton* grass
[54]*In* Greek mythology, Hippomenes won the swift Atalanta's hand by defeating her in a footrace. At the advice of Aphrodite, Hippomenes slowed Atalanta by dropping the three apples of the Hesperides.

meteors, earth and sea, minerals, plants, animals,—but much more of nature under constraint and vexed; that is to say, when by art and the hand of man she is forced out of her natural state, and squeezed and moulded. Therefore I set down at length all experiments of the mechanical arts, of the operative part of the liberal arts, of the many crafts which have not yet grown into arts properly so called, so far as I have been able to examine them and as they conduce to the end in view. Nay (to say the plain truth) I do in fact (low and vulgar as men may think it) count more upon this part both for helps and safeguards than upon the other; seeing that the nature of things betrays itself more readily under the vexations of art than in its natural freedom.

Nor do I confine the history to Bodies; but I have thought it my duty besides to make a separate history of such Virtues[55] as may be considered cardinal in nature. I mean those original passions[56] or desires[57] of matter which constitute the primary elements of nature; such as Dense and Rare, Hot and Cold, Solid and Fluid, Heavy and Light, and several others.[58]

Then again, to speak of subtlety: I seek out and get together a kind of experiments much subtler and simpler than those which occur accidentally.[59] For I drag into light many things which no one who was not proceeding by a regular and certain way to the discovery of causes would have thought of inquiring after; being indeed in themselves of no great use; which shows that they were not sought for on their own account; but having just the same relation to things and works which the letters of the alphabet have to speech and words—which, though in themselves useless, are the elements of which all discourse is made up.

Further, in the selection of the relations and experiments I conceive I have been a more cautious purveyor

[55]*Qualities,* especially qualities that are beneficial to the human body
[56]*The* way a thing may be affected by an external agency
[57]*Tendencies*
[58]*See* Anderson, *The Philosophy of Francis Bacon,* pp. 34–35.
[59]*Casually*

than those who have hitherto dealt with natural history. For I admit nothing but on the faith of eyes, or at least of careful and severe examination; so that nothing is exaggerated for wonder's sake, but what I state is sound and without mixture of fables or vanity. All received or current falsehoods also (which by strange negligence have been allowed for many ages to prevail and become established) I proscribe and brand by name; that the sciences may be no more troubled with them. For it has been well observed that the fables and superstitions and follies which nurses instil into children do serious injury to their minds; and the same consideration makes me anxious, having the management of the childhood as it were of philosophy in its course of natural history, not to let it accustom itself in the beginning to any vanity. Moreover, whenever I come to a new experiment of any subtlety (though it be in my own opinion certain and approved), I nevertheless subjoin a clear account of the manner in which I made it; that men knowing exactly how each point was made out, may see whether there be any error connected with it, and may arouse themselves to devise proofs more trustworthy and exquisite, if such can be found; and finally, I interpose everywhere admonitions and scruples and cautions, with a religious care to eject, repress, and as it were exorcise every kind of phantasm.

Lastly, knowing how much the sight of man's mind is distracted by experience and history, and how hard it is at the first (especially for minds either tender or preoccupied) to become familiar with nature, I not unfrequently subjoin observations of my own, being as the first offers, inclinations, and as it were glances of history towards philosophy; both by way of an assurance to men that they will not be kept for ever tossing on the waves of experience, and also that when the time comes for the intellect to begin its work, it may find everything the more ready. By such a natural history then as I have described, I conceive that a safe and convenient approach may be made to nature, and matter supplied of good quality and well prepared for the understanding to work upon.

And now that we have surrounded the intellect with faithful helps and guards, and got together with most careful selection a regular army of divine works, it may seem that we have no more to do but to proceed to philosophy itself. And yet in a matter so difficult and doubtful there are still some things which it seems necessary to premise,[60] partly for convenience of explanation, partly for present use.

Of these the first is to set forth examples of inquiry and invention according to my method, exhibited by anticipation in some particular subjects; choosing such subjects as are at once the most noble in themselves among those under inquiry, and most different one from another; that there may be an example in every kind. I do not speak of those examples which are joined to the several precepts and rules by way of illustration (for of these I have given plenty in the second part of the work); but I mean actual types and models, by which the entire process of the mind and the whole fabric and order of invention from the beginning to the end, in certain subjects, and those various and remarkable, should be set as it were before the eyes. For I remember that in the mathematics it is easy to follow the demonstration when you have a machine beside you; whereas without that help all appears involved and more subtle than it really is. To examples of this kind,—being in fact nothing more than an application of the second part in detail and at large,—the fourth part[61] of the work is devoted.

The fifth part[62] is for temporary use only, pending the completion of the rest; like interest payable from time to time until the principal be forthcoming. For I do not make so blindly for the end of my journey, as to neglect anything useful that may turn up by the way. And therefore I include in this fifth part such things as I have myself discovered, proved, or added,—not however according to the true rules and methods of interpretation, but by the ordinary use of the understanding in inquiring

[60]*To* set forth beforehand
[61]*Bacon* completed only a very small fragment of the fourth part.
[62]*Bacon* completed only a very small fragment of the fifth part.

and discovering. For besides that I hope my speculations may in virtue of my continual conversancy with nature have a value beyond the pretensions of my wit, they will serve in the meantime for wayside inns, in which the mind may rest and refresh itself on its journey to more certain conclusions. Nevertheless I wish it to be understood in the meantime that they are conclusions by which (as not being discovered and proved by the true form of interpretation) I do not at all mean to bind myself. Nor need any one be alarmed at such suspension of judgment, in one who maintains not simply that nothing can be known, but only that nothing can be known except in a certain course and way; and yet establishes provisionally certain degrees of assurance, for use and relief until the mind shall arrive at a knowledge of causes in which it can rest. For even those schools of philosophy which held the absolute impossibility of knowing anything were not inferior to those which took upon them to pronounce. But then they did not provide helps for the sense and understanding, as I have done, but simply took away all their authority: which is quite a different thing—almost the reverse.

The sixth part[63] of my work (to which the rest is subservient and ministrant)[64] discloses and sets forth that philosophy which by the legitimate, chaste, and severe course of inquiry which I have explained and provided is at length developed and established. The completion however of this last part is a thing both above my strength and beyond my hopes. I have made a beginning of the work—a beginning, as I hope, not unimportant:—the fortune of the human race will give the issue;—such an issue, it may be, as in the present condition of things and men's minds cannot easily be conceived or imagined. For the matter in hand is no mere felicity[65] of speculation, but the real business and fortunes of the human race, and all power of operation. For man is but the

[63]*As* a formally identified part of the plan, the sixth part does not exist in Bacon's writing. See pp. x–xiii, above.
[64]*A* minister to
[65]*Happy* expression

servant and interpreter of nature: what he does and what he knows is only what he has observed of nature's order in fact or in thought; beyond this he knows nothing and can do nothing. For the chain of causes cannot by any force be loosed or broken, nor can nature be commanded[66] except by being obeyed. And so those twin objects, human Knowledge and human Power, do really meet in one; and it is from ignorance of causes that operation fails.

And all depends on keeping the eye steadily fixed upon the facts of nature and so receiving their images simply as they are. For God forbid that we should give out a dream of our own imagination for a pattern of the world; rather may he graciously grant to us to write an apocalypse[67] or true vision of the footsteps of the Creator imprinted on his creatures.

Therefore do thou, O Father, who gavest the visible light as the first fruits of creation, and didst breathe into the face of man the intellectual light as the crown and consummation thereof, guard and protect this work, which coming from thy goodness returneth to thy glory. Thou when thou turnedst to look upon the works which thy hands had made, sawest that all was very good, and didst rest from thy labours. But man, when he turned to look upon the work which his hands had made, saw that all was vanity and vexation of spirit, and could find no rest therein. Wherefore if we labour in thy works with the sweat of our brows thou wilt make us partakers of thy vision and thy sabbath. Humbly we pray that this mind may be steadfast in us, and that through these our hands, and the hands of others to whom thou shalt give the same spirit, thou wilt vouchsafe to endow the human family with new mercies.

[66]*The* Latin is *vincitur*, which should be translated "conquered."
[67]*Prophetic* revelation

The

FIRST PART OF THE INSTAURATION,

Which comprises the

DIVISIONS OF THE SCIENCES,

IS WANTING.

But some account of them will be found in the
Second
Book of the
"Proficience and Advancement of Learning,
Divine and Human."

Next comes

The

SECOND PART OF THE INSTAURATION,

WHICH EXHIBITS

THE ART ITSELF OF INTERPRETING NATURE,

AND OF THE TRUER EXERCISE OF THE INTELLECT;

Not however in the form of a regular Treatise, but only a
Summary
digested into Aphorisms.

NEW ATLANTIS:
A WORK UNFINISHED.

WRITTEN BY

THE RIGHT HONOURABLE

FRANCIS LORD VERULAM, VISCOUNT ST. ALBAN.[1]

[1]*Bacon's* titles. He was also knighted.

TO THE READER.

This fable my Lord devised, to the end that he might exhibit therein a model or description of a college instituted for the interpreting of nature and the producing of great and marvellous works for the benefit of men, under the name of Salomon's House, or the College of the Six Days' Works. And even so far his Lordship hath proceeded, as to finish that part. Certainly the model is more vast and high than can possibly be imitated in all things; notwithstanding most things therein are within men's power to effect. His Lordship thought also in this present fable to have composed a frame of Laws, or of the best state or mould of a commonwealth; but foreseeing it would be a long work, his desire of collecting the Natural History diverted him, which he preferred many degrees before it.

This work of the *New Atlantis* (as much as concerneth the English edition) his Lordship designed for this place[2]; in regard it hath so near affinity (in one part of it) with the preceding Natural History.

W. RAWLEY.[3]

[2]*The New Atlantis* was published in 1627, after Bacon's death. It appeared at the end of the volume containing the *Sylva Sylvarum* ("A Forest of Materials"), a major part of his natural history.
[3]*Bacon's* secretary

NEW ATLANTIS

We sailed from Peru, (where we had continued by the space of[4] one whole year,) for China and Japan, by the South Sea[5]; taking with us victuals for twelve months; and had good winds from the east, though soft and weak, for five months' space and more. But then the wind came about,[6] and settled in the west for many days, so as we could make little or no way, and were sometimes in purpose[7] to turn back. But then again there arose strong and great winds from the south, with a point east; which carried us up (for all that we could do) towards the north: by which time our victuals failed us, though we had made good spare of them.[8] So that finding ourselves in the midst of the greatest wilderness of waters in the world, without victual, we gave ourselves for lost men, and prepared for death. Yet we did lift up our hearts and voices to God above, who *showeth his wonders in the deep*;[9] beseeching him of his mercy, that as in the beginning he discovered[10] the face of the deep, and brought forth dry land, so he would now discover land to us, that we might not perish. And it came to pass that the next day about evening, we saw within a kenning[11] before us, towards the north, as it were thick clouds, which did put us in some hope of land; knowing how that part of the South Sea was utterly unknown; and might have islands or continents, that hitherto were not come to

[4]*Stayed* for
[5]*Pacific* Ocean
[6]*Came* from the opposite direction
[7]*Decided,* resolved
[8]*Conserved* them
[9]*Psalms* 107: 23–32
[10]*Uncovered,* drew back
[11]*Distance* measured by the range of sight: 20 miles

light. Wherefore we bent[12] our course thither, where we saw the appearance of land, all that night; and in the dawning of the next day, we might plainly discern that it was a land; flat to our sight, and full of boscage;[13] which made it shew the more dark. And after an hour and a half's sailing, we entered into a good haven, being the port of a fair city; not great indeed, but well built, and that gave a pleasant view from the sea: and we thinking every minute long till we were on land, came close to the shore, and offered[14] to land. But straightways we saw divers[15] of the people, with bastons[16] in their hands, as it were forbidding us to land; yet without any cries or fierceness, but only as warning us off by signs that they made. Whereupon being not a little discomforted, we were advising with ourselves what we should do. During which time there made forth to us a small boat, with about eight persons in it; whereof one of them had in his hand a tipstaff[17] of a yellow cane, tipped at both ends with blue, who came aboard our ship, without any show of distrust at all. And when he saw one of our number present himself somewhat afore the rest, he drew forth a little scroll of parchment, (somewhat yellower than our parchment, and shining like the leaves of writing tables,[18] but otherwise soft and flexible,) and delivered it to our foremost man. In which scroll were written in ancient Hebrew, and in ancient Greek, and in good Latin of the School,[19] and in Spanish, these words; "Land ye not, none[20] of you; and provide to be gone from this coast within sixteen days, except you have further time given you. Meanwhile, if you want fresh water, or victual, or help for your sick, or that your ship needeth repair, write down your wants, and you shall have that

[12]*Turned*

[13]*Forest*

[14]*Showed* intention to

[15]*Various* (not people who dive!)

[16]*Batons,* truncheons

[17]*A* staff with a tip or cap of metal carried as a badge by certain officials

[18]*Tablets* for inscriptions

[19]*Latin* as written by the Scholastic philosophers, university Latin

[20]*An* acceptable double negative in Bacon's time

which belongeth to mercy." This scroll was signed with
a stamp of cherubins'[21] wings, not spread but hanging
downwards, and by them a cross. This being delivered,
the officer returned, and left only a servant with us to
receive our answer. Consulting hereupon amongst ourselves, we were much perplexed. The denial of landing
and hasty warning us away troubled us much; on the
other side, to find that the people had languages[22] and
were so full of humanity, did comfort us not a little.
And above all, the sign of the cross to that instrument[23]
was to us a great rejoicing, and as it were a certain
presage of good. Our answer was in the Spanish tongue;
"That for our ship, it was well; for we had rather met
with calms and contrary winds than any tempests. For
our sick, they were many, and in very ill case;[24] so that
if they were not permitted to land, they ran danger of
their lives." Our other wants we set down in particular;
adding, "that we had some little store of merchandise,
which if it pleased them to deal for, it might supply our
wants without being chargeable unto them."[25] We offered some reward in pistolets[26] unto the servant, and
a piece of crimson velvet to be presented to the officer;
but the servant took them not, nor would scarce look
upon them; and so left us, and went back in another
little boat which was sent for him.

About three hours after we had dispatched our answer,
there came towards us a person (as it seemed) of place.[27]
He had on him a gown with wide sleeves, of a kind of
water chamolet,[28] of an excellent azure colour, far more
glossy than ours; his under apparel was green; and so
was his hat, being in the form of a turban, daintily made,
and not so huge as the Turkish turbans; and the locks
of his hair came down below the brims of it. A reverend
man was he to behold. He came in a boat, gilt in some

[21]*A* biblical figure with wings, a human head, and an animal body
[22]*Were* conversant in European languages
[23]*The* scroll
[24]*Poor* condition
[25]*Without* costing them anything
[26]*Gold* coins
[27]*High* rank
[28]*A* costly oriental fabric

part of it, with four persons more only in that boat; and was followed by another boat, wherein were some twenty. When he was come within a flight-shot[29] of our ship, signs were made to us that we should send forth some to meet him upon the water; which we presently did in our ship-boat,[30] sending the principal man amongst us save one,[31] and four of our number with him. When we were come within six yards of their boat, they called to us to stay,[32] and not to approach farther; which we did. And thereupon the man whom I before described stood up, and with a loud voice in Spanish, asked, "Are ye Christians?" We answered, "We were;" fearing the less, because of the cross we had seen in the subscription. At which answer the said person lifted up his right hand towards heaven, and drew it softly to his mouth, (which is the gesture they use when they thank God,) and then said: "If ye will swear (all of you) by the merits of the Saviour that ye are no pirates, nor have shed blood lawfully nor unlawfully within forty days past,[33] you may have licence to come on land." We said, "We were all ready to take that oath." Whereupon one of those that were with him, being (as it seemed) a notary, made an entry of this act. Which done, another of the attendants of the great person, which was with him in the same boat, after his lord had spoken a little to him, said aloud; "My lord would have you know, that it is not of pride or greatness that he cometh not aboard your ship; but for that in your answer you declare that you have many sick amongst you, he was warned by the Conservator of Health of the city that he should keep a distance." We bowed ourselves towards him, and answered, "We were his humble servants; and accounted for great honour and singular humanity towards us that which was already done; but hoped well that the nature of the sickness of our men was not infectious." So he returned; and a while

---

[29]*The* flight distance of a long-range arrow
[30]*Small* boat carried or towed by a ship
[31]*Second* in command
[32]*Stop*
[33]*See* Exodus 20: 13; 21: 12–14; Numbers 35: 9–34; Deut. 5: 17; cf. Maimonides, *Guide of the Perplexed*, III, 40, 41.

after came the notary to us aboard our ship; holding in
his hand a fruit of that country, like an orange, but of
color between orange-tawney and scarlet, which cast a
most excellent odour. He used it (as it seemeth) for a
preservative against infection. He gave us our oath; "By
the name of Jesus and his merits:" and after told us that
the next day by six of the clock in the morning we should
be sent to, and brought to the Strangers' House, (so he
called it,) where we should be accomodated of things
both for our whole[34] and for our sick. So he left us; and
when we offered him some pistolets, he smiling said,
"He must not be twice paid for one labour:" meaning
(as I take it) that he had salary sufficient of the state
for his service. For (as I after learned) they call an officer
that taketh rewards, *twice paid.*

The next morning early, there came to us the same
officer that came to us at first with his cane, and told
us, "He came to conduct us to the Strangers' House;
and that he had prevented the hour,[35] because we might
have the whole day before us for our business. "For,"
said he, "if you will follow my advice, there shall first
go with me some few of you, and see the place, and
how it may be made convenient for you; and then you
may send for your sick, and the rest of your number
which ye will bring on land." We thanked him, and said,
"That this care which he took of desolate strangers God
would reward." And so six of us went on land with him:
and when we were on land, he went before us, and
turned to us, and said, "He was but our servant, and
our guide." He led us through three fair streets; and all
the way we went there were gathered some people on
both sides standing in a row; but in so civil a fashion,
as if it had been not to wonder at us but to welcome
us; and divers of them, as we passed by them, put their
arms a little abroad; which is their gesture when they
bid any welcome. The Strangers' House is a fair and
spacious house, built of brick, of somewhat a bluer colour
than our brick; and with handsome windows, some of

[34]*Well*
[35]*He* had come early

glass, some of a kind of cambric[36] oiled. He brought us first into a fair parlour above stairs, and then asked us, "What number of persons we were? And how many sick?" We answered, "We were in all (sick and whole) one and fifty persons, whereof our sick were seventeen." He desired us to have patience a little, and to stay till he came back to us; which was about an hour after; and then he led us to see the chambers which were provided for us, being in number nineteen: they having cast it[37] (as it seemeth) that four of those chambers, which were better than the rest, might receive four of the principal men of our company, and lodge them alone by themselves; and the other fifteen chambers were to lodge us two and two together.[38] The chambers were handsome and cheerful chambers, and furnished civilly. Then he led us to a long gallery, like a dorture,[39] where he showed us all along the one side (for the other side was but wall and window) seventeen cells, very neat ones, having partitions of cedar wood. Which gallery and cells, being in all forty, (many more than we needed,) were instituted as an infirmary for sick persons. And he told us withal,[40] that as any of our sick waxed[41] well, he might be removed from his cell to a chamber; for which purpose there were set forth ten spare chambers, besides the number we spake of before. This done, he brought us back to the parlour, and lifting up his cane a little, (as they do when they give any charge or command,) said to us, "Ye are to know that the custom of the land requireth, that after this day and to-morrow, (which we give you for removing of your people from your ship,) you are to keep within doors for three days. But let it not trouble you, nor do not think yourselves restrained, but rather left to your rest and ease. You shall want nothing, and there are six of our people appointed to attend you, for any business you may have abroad." We gave him thanks with all

[36]**A** kind of white linen
[37]*Reckoned*
[38]*By* twos
[39]*Dormitory*
[40]*Moreover*
[41]*Became*

affection and respect, and said, "God surely is manifested in this land." We offered him also twenty pistolets; but he smiled, and only said; "What? twice paid!" And so he left us. Soon after our dinner was served in; which was right good viands,[42] both for bread and meat: better than any collegiate diet[43] that I have known in Europe. We had also drink of three sorts, all wholesome and good; wine of the grape; a drink of grain, such as is with us our ale, but more clear; and a kind of cider made of a fruit of that country; a wonderful pleasing and refreshing drink. Besides, there were brought in to us great store of those scarlet oranges for our sick; which (they said) were an assured remedy for sickness taken at sea. There was given us also a box of small grey or whitish pills, which they wished our sick should take, one of the pills every night before sleep; which (they said) would hasten their recovery. The next day, after that our trouble of carriage and removing of our men and goods out of our ship was somewhat settled and quiet, I thought good to call our company together; and when they were assembled said unto them; "My dear friends, let us know ourselves, and how it standeth with us. We are men cast on land, as Jonas[44] was out of the whale's belly, when we were as buried in the deep: and now we are on land, we are but between death and life; for we are beyond both the old world and the new; and whether ever we shall see Europe, God only knoweth. It is a kind of miracle hath brought us hither: and it must be little less that shall bring us hence.[45] Therefore in regard of our deliverance past, and our danger present and to come, let us look up to God, and every man reform his own ways. Besides we are come here amongst a Christian people, full of piety and humanity: let us not bring that confusion of face[46] upon ourselves, as to show our vices or unworthiness before them. Yet there is more. For they have by commandment (though in form of courtesy)

[42]*Food*
[43]*Food* served for an organized society
[44]*Jonah.* See Jonah 1–4.
[45]*From* this place
[46]*Shame*

cloistered us within these walls for three days: who knoweth whether it be not to take some taste[47] of our manners and conditions? and if they find them bad, to banish us straightways; if good, to give us further time. For these men that they have given us for attendance may withal have an eye upon us. Therefore for God's love, and as we love the weal[48] of our souls and bodies, let us so behave ourselves as we may be at peace with God, and may find grace in the eyes of this people." Our company with one voice thanked me for my good admonition, and promised me to live soberly and civilly, and without giving any the least occasion of offence. So we spent our three days joyfully and without care, in expectation what would be done with us when they were expired. During which time, we had every hour joy of the amendment[49] of our sick; who thought themselves cast into some divine pool of healing,[50] they mended so kindly[51] and so fast.

The morrow[52] after our three days were past, there came to us a new man that we had not seen before, clothed in blue as the former was, save that his turban was white, with a small red cross on the top. He had also a tippet[53] of fine linen. At his coming in, he did bend to us a little, and put his arms abroad. We of our parts saluted him in a very lowly and submissive manner; as looking that from him we should receive sentence of life or death. He desired to speak with some few of us: whereupon six of us only stayed, and the rest avoided[54] the room. He said, "I am by office governor of this House of Strangers, and by vocation I am a Christian priest; and therefore am come to you to offer you my service, both as strangers and chiefly as Christians. Some things I may tell you, which I think you will not be unwilling to hear. The state hath given you licence to stay on land

[47]*To* observe
[48]*Well-being*
[49]*Recovery*
[50]*Cf.* John 5: 2–4.
[51]*Naturally*
[52]*Morning*
[53]*Long* scarf or cape
[54]*Left*

for the space of six weeks: and let it not trouble you if your occasions ask[55] further time, for the law in this point is not precise; and I do not doubt but myself shall be able to obtain for you such further time as many be convenient. Ye shall also understand, that the Strangers' House is at this time rich, and much aforehand;[56] for it hath laid up revenue these thirty-seven years; for so long it is since any stranger arrived in this part: and therefore take ye no care; the state will defray you[57] all the time you stay; neither shall you stay one day the less for that. As for any merchandise ye have brought, ye shall be well used,[58] and have your return either in merchandise or in gold and silver: for to us it is all one. And if you have any other request to make, hide it not. For ye shall find we will not make your countenance to fall[59] by the answer ye shall receive. Only this I must tell you, that none of you must go above a *karan"*[60] (that is with them a mile and an half) "from the walls of the city, without especial leave." We answered, after we had looked awhile one upon another, admiring this gracious and parent-like usage;[61] "That we could not tell what to say: for we wanted[62] words to express our thanks; and his noble free offers left us nothing to ask. It seemed to us that we had before us a picture of our salvation in heaven; for we that were awhile since[63] in the jaws of death, were now brought into a place where we found nothing but consolations. For the commandment laid upon us, we would not fail to obey it, though it was impossible but our hearts should be inflamed to tread further upon this happy and holy ground." We added; "That our ton-

[55]*Your* needs require
[56]*Prepared* or provided for the future
[57]*Pay* your expenses
[58]*Well* treated, treated fairly
[59]*We* will not disappoint you
[60]*This* comes from the Hebrew word *keren,* which means "horn" and is used to symbolize people's strength. It is used to refer to a ruler of the Davidic line; see Psalms 132: 17; Ezekiel 29: 21. The verb form means to send out rays or display horns.
[61]*Treatment*
[62]*Lacked*
[63]*Just* before

gues should first cleave to the roofs of our mouths,[64] ere we should forget either his reverend person or this whole nation in our prayers." We also most humbly besought him to accept of us as his true servants, by as just a right as ever men on earth were bounden; laying and presenting both our persons and all we had at his feet. He said; "He was a priest, and looked for a priest's reward: which was our brotherly love and the good of our souls and bodies." So he went from us, not without tears of tenderness in his eyes; and left us also confused[65] with joy and kindness, saying amongst ourselves, "That we were come into a land of angels, which did appear to us daily and prevent us[66] with comforts, which we thought not of, much less expected."

The next day, about ten of the clock, the governor came to us again, and after salutations said familiarly, "That he was come to visit us": and called for a chair, and sat him down: and we, being some ten of us, (the rest were of the meaner sort,[67] or else gone abroad,) sat down with him. And when we were set, he began thus: "We of this island of Bensalem,"[68] (for so they call it in their language,) "have this; that by means of our solitary situation, and of the laws of secrecy which we have for our travellers, and our rare admission of strangers, we know well most part of the habitable world, and are ourselves unknown. Therefore because he that knoweth least is fittest to ask questions, it is more reason, for the entertainment of the time,[69] that ye ask me questions, than that I ask you." We answered; "That we humbly thanked him that he would give us leave so to do: and that we conceived by the taste we had already, that there was no worldly thing on earth more worthy to be known than the state of that happy land. But above all," (we said,) "since that we were met from the several ends of

[64]*Psalms* 137: 6
[65]*Amazed*
[66]*Anticipate* our needs
[67]*Of* a humbler class
[68]*"Bensalem"* is a combination of Hebrew words (*ben, shalem*) meaning "son or offspring of peace, safety, and completeness."
[69]*For* passing the time

the world, and hoped assuredly that we should meet one day in the kingdom of heaven, (for that we were both parts Christians,) we desired to know (in respect that land was so remote, and so divided by vast and unknown seas, from the land where our Saviour walked on earth,) who was the apostle[70] of that nation, and how it was converted to the faith?" It appeared in his face that he took great contentment in this our question: he said, "Ye knit my heart to you, by asking this question in the first place; for it sheweth that you *first seek the kingdom of heaven*;[71] and I shall gladly and briefly satisfy your demand.

"About twenty years after the ascension of our Saviour, it came to pass that there was seen by the people of Renfusa,[72] (a city upon the eastern coast of our island,)[73] within night, (the night was cloudy and calm,) as it might be some mile into the sea,[74] a great pillar of light; not sharp, but in form of a column or cylinder, rising from the sea a great way up towards heaven: and on the top of it was seen a large cross of light, more bright and resplendent than the body of the pillar. Upon which so strange a spectacle, the people of the city gathered apace[75] together upon the sands, to wonder; and so after put themselves into a number of small boats, to go nearer to this marvellous sight. But when the boats were come within about sixty yards of the pillar, they found themselves all bound, and could go no further; yet so as they might move to go about,[76] but might not approach nearer: so as the boats stood all as in a theatre, beholding this light as an heavenly sign. It so fell out,[77] that there was in one of the boats one of the wise men

[70]*The* missionary who first plants Christianity in a region
[71]*Matt.* 6: 33
[72]*"Renfusa"* is a combination of the Greek words *rhen* and *phusis* meaning "sheep-natured" or "sheep-like."
[73]*Since* the wind bringing the sailors to the island was from the south with "a point east," the sailors landed on the east side of the island.
[74]*A* mile out to sea
[75]*Quickly*
[76]*To* turn around to go back
[77]*It* so happened

of the society of Salomon's House; which house or college (my good brethren) is the very eye[78] of this kingdom; who having awhile attentively and devoutly viewed and contemplated this pillar and cross, fell down upon his face; and then raised himself upon his knees, and lifting up his hands to heaven, made his prayers in this manner:

" 'Lord God of heaven and earth, thou hast vouchsafed of thy grace to those of our order, to know thy works of creation, and the secrets of them; and to discern (as far as appertaineth to the generations of men) between divine miracles, works of nature, works of art, and impostures and illusions of all sorts. I do here acknowledge and testify before this people, that the thing which we now see before our eyes is thy Finger and a true Miracle; and forasmuch as[79] we learn in our books that thou never workest miracles but to a divine and excellent end, (for the laws of nature are thine own laws, and thou exceedest them not but upon great cause,) we most humbly beseech thee to prosper[80] this great sign, and to give us the interpretation and use of it in mercy; which thou dost in some part secretly promise by sending it unto us.'

"When he had made his prayer, he presently found the boat he was in moveable and unbound; whereas all the rest remained still fast; and taking that for an assurance of leave to approach, he caused the boat to be softly and with silence rowed towards the pillar. But ere he came near it, the pillar and cross of light brake up, and cast itself abroad, as it were, into a firmament of many stars; which also vanished soon after, and there was nothing left to be seen but a small ark or chest of cedar, dry, and not wet at all with water, though it swam. And in the fore-end of it, which was towards him, grew a small green branch of palm; and when the wise man had taken it with all reverence into his boat, it opened of itself, and there were found in it a Book and a Letter; both written in fine parchment, and wrapped in sindons

[78]*Seat* of intelligence
[79]*Since*
[80]*To* cause to succeed

of linen.[81] The Book contained all the canonical books[82] of the Old and New Testament, according as you have them, (for we know well what the Churches with you receive); and the Apocalypse[83] itself, and some other books of the New Testament which were not at that time written,[84] were nevertheless in the Book. And for the Letter, it was in these words:

" 'I Bartholomew,[85] a servant of the Highest, and Apostle of Jesus Christ, was warned by an angel that appeared to me in a vision of glory, that I should commit this ark to the floods of the sea. Therefore I do testify and declare unto that people where God shall ordain this ark to come to land, that in the same day is come unto them salvation and peace and goodwill, from the Father, and from the Lord Jesus.'

"There was also in both these writings, as well the Book as the Letter, wrought a great miracle, conform to that of the Apostles in the original Gift of Tongues.[86] For there being at that time in this land Hebrews, Persians, and Indians, besides the natives, every one read upon the Book and Letter, as if they had been written in his own language. And thus was this land saved from infidelity (as the remain of the old world was from water) by an ark, through the apostolical and miraculous evangelism of St. Bartholomew." And here he paused, and a messenger came, and called him from us. So this was all that passed in that conference.

The next day, the same governor came again to us immediately after dinner, and excused himself, saying, "That the day before he was called from us somewhat abruptly, but now he would make us amends, and spend time with us, if we held his company and conference

[81]*Wrappings* of fine, thin linen
[82]*The* canonical books of the Bible are those accepted as Holy Scripture. The Apocrypha are those books included in the Septuagint (Greek Bible) and the Vulgate (Latin Bible) but excluded from the Jewish and Protestant canons of the Old Testament.
[83]*The* Revelation to John
[84]*At* least Acts, Paul's Epistles, and the Revelation to John!
[85]*St.* Bartholomew was the man said to have preached the gospel to India; i.e., the islands off of the southern coast of Asia.
[86]*Acts* 2: 1–16

agreeable." We answered, "That we held it so agreeable
and pleasing to us, as we forgot both dangers past and
fears to come, for the time we heard him speak; and
that we thought an hour spent with him, was worth years
of our former life." He bowed himself a little to us, and
after we were set again, he said; "Well, the questions
are on your part." One of our number said, after a little
pause; "That there was a matter we were no less desirous
to know, than fearful to ask, lest we might presume too
far. But encouraged by his rare[87] humanity towards us,
(that could scarce think ourselves strangers, being his
vowed and professed servants,) we would take the
hardiness[88] to propound it: humbly beseeching him, if
he thought it not fit to be answered, that he would
pardon it, though he rejected it." We said; "We well
observed those his words, which he formerly spake, that
this happy island where we now stood was known to
few, and yet knew most of the nations of the world;
which we found to be true, considering they had the
languages of Europe, and knew much of our state[89] and
business; and yet we in Europe (notwithstanding all the
remote discoveries and navigations of this last age,) never
heard any of the least inkling or glimpse of this island.
This we found wonderful strange; for that all nations
have inter-knowledge one of another either by voyage
into foreign parts, or by strangers that come to them:
and though the traveller into a foreign country doth
commonly know more by the eye, than he that stayeth
at home can by relation of the traveller; yet both ways
suffice to make a mutual knowledge, in some degree,
on both parts. But for this island, we never heard tell
of any ship of theirs that had been seen to arrive upon
any shore of Europe; no, nor of either the East or West
Indies;[90] nor yet of any ship of any other part of the
world that had made return from them. And yet the

[87]*Unusual,* remarkably fine
[88]*Be* so bold as to
[89]*Condition*
[90]*East Indies:* India and adjacent regions and islands; West Indies:
lands of Western Hemisphere discovered in the 15th and 16th cen-
turies

marvel rested not in this. For the situation[91] of it (as his lordship said) in the secret conclave[92] of such a vast sea might cause it. But then that they should have knowledge of the languages, books, affairs, of those that lie such a distance from them, it was a thing we could not tell what to make of; for that it seemed to us a condition and propriety[93] of divine powers and beings, to be hidden and unseen to others, and yet to have others open and as in a light to them.'' At this speech the governor gave a gracious smile, and said; ''That we did well to ask pardon for this question we now asked; for that it imported as if[94] we thought this land a land of magicians, that sent forth spirits of the air into all parts, to bring them news and intelligence of other countries.'' It was answered by us all, in all possible humbleness, but yet with a countenance taking knowledge[95] that we knew that he spake it but merrily, ''That we were apt enough to think there was somewhat supernatural in this island; but yet rather as angelical than magical. But to let his lordship know truly what it was that made us tender[96] and doubtful[97] to ask this question, it was not any such conceit, but because we remembered he had given a touch in his former speech, that this land had laws of secrecy touching strangers.'' To this he said; ''You remember it aright; and therefore in that I shall say to you I must reserve some particulars, which it is not lawful for me to reveal; but there will be enough left to give you satisfaction.

''You shall understand (that which perhaps you will scarce think credible) that about three thousand years ago, or somewhat more, the navigation of the world, (specially for remote voyages,) was greater than at this day. Do not think with yourselves that I know not how much it is increased with you within these six-score

[91]*Location*
[92]*Private* room, inner chamber, closet
[93]*Special* quality
[94]*Implied,* signified that
[95]*A* gesture showing
[96]*Cautious*
[97]*Apprehensive*

years:[98] I know it well: and yet I say greater then than now; whether it was, that the example of the ark, that saved the remnant of men from the universal deluge, gave men confidence to adventure upon the waters; or what it was; but such is the truth. The Phoenicians,[99] and especially the Tyrians,[100] had great fleets. So had the Carthaginians,[101] their colony, which is yet further west. Toward the east, the shipping of Egypt and of Palestina[102] was likewise great. China also, and the great Atlantis (that you call America), which have now but junks and canoes, abounded then in tall ships. This island (as appeareth by faithful registers of those times) had then fifteen hundred strong ships, of great content.[103] Of all this there is with you sparing memory, or none; but we have large knowledge thereof.

"At that time, this land was known and frequented by the ships and vessels of all the nations before named. And (as it cometh to pass) they had many times men of other countries, that were no sailors, that came with them; as Persians, Chaldeans,[104] Arabians; so as almost all nations of might and fame resorted hither; of whom we have some stirps[105] and little tribes with us at this day. And for our own ships, they went sundry voyages, as well to your Straits, which you call the Pillars of Hercules, as to other parts in the Atlantic and Mediterrane Seas; as to Paguin (which is the same with Cambaline) and Quinzy, upon the Oriental Seas, as far as to the borders of the East Tartary.[106]

[98]*Taking* the year 1492 as the date of the opening of the New World to European navigation, the date of the sailors' voyage is six-score (120) years later; *i.e.*, 1612.

[99]*Inhabitants* of the ancient country north of Palestine on the Syrian coast

[100]*Inhabitants* of the ancient Phoenician city of Tyre

[101]*Phoenician* colonists of the ancient city of North Africa, now Tunis

[102]*Palestine*

[103]*Tonnage*, capacity

[104]*Inhabitants* of ancient Mesopotamia

[105]*Lineages*

[106]*Pillars* of Hercules: the Straits of Gibraltar; Paguin, Cambaline: names for Peking; Quinzy: Hangchow; Borders of East Tartary: East Coast of Asia, north of China

"At the same time, and an age after, or more, the inhabitants of the great Atlantis did flourish. For though the narration and description which is made by a great man[107] with you, that the descendants of Neptune[108] planted[109] there; and of the magnificent temple, palace, city, and hill; and the manifold streams of goodly navigable rivers, (which, as so many chains, environed the same site and temple); and the several degrees[110] of ascent whereby men did climb up to the same, as if it had been a *scala coeli*;[111] be all poetical and fabulous: yet so much is true, that the said country of Atlantis, as well that of Peru, then called Coya, as that of Mexico,[112] then named Tyrambel, were mighty and proud kingdoms in arms, shipping, and riches: so mighty, as at one time (or at least within the space of ten years) they both made two great expeditions; they of Tyrambel through the Atlantic to the Mediterrane Sea; and they of Coya through the South Sea upon this our island. And for the former of these, which was into Europe, the same author amongst you (as it seemeth) had some relation from the Egyptian priest whom he citeth.[113] For assuredly such a thing there was. But whether it were the ancient Athenians that had the glory of the repulse and resistance of those forces, I can say nothing: but certain it is, there never came back either ship or man from that voyage. Neither had the other voyage of those of Coya upon us had better fortune,[114] if they had not met with enemies of greater clemency. For the king of this island (by name Altabin)[115] a wise man and a great warrior, knowing well both his own strength and that of his enemies, handled the matter so, as he cut off their land-forces from their ships; and

---

[107]*Plato Timaeus* 21e1–25d7; *Critias* 113a1–121c4
[108]*God* of the sea in Roman mythology corresponding to the Greek god Poseidon
[109]*Settled*
[110]*Steps*
[111]*Ladder* of heaven
[112]*The* ancient Atlantis was apparently in North America; Coya and Tyrambel were in Central and South America.
[113]*Timaeus* 21a7 ff.; *Critias* 110a6 ff.
[114]*The* Coyans would have fared no better
[115]*Combination* of Latin words meaning "twice lofty"

entoiled[116] both their navy and their camp with a greater power than theirs, both by sea and land; and compelled them to render themselves without striking stroke: and after they were at his mercy, contenting himself only with their oath that they should no more bear arms against him, dismissed them all in safety. But the Divine Revenge overtook not long after those proud enterprises. For within less than the space of one hundred years, the great Atlantis was utterly lost and destroyed: not by a great earthquake, as your man saith,[117] (for that whole tract is little subject to earthquakes,) but by a particular deluge or inundation;[118] those countries having, at this day, far greater rivers and far higher mountains to pour down waters, than any part of the old world.[119] But it is true that the same inundation was not deep; not past forty foot, in most places, from the ground: so that although it destroyed man and beast generally, yet some few wild inhabitants of the wood escaped. Birds also were saved by flying to the high trees and woods. For as for men, although they had buildings in many places higher than the depth of the water, yet that inundation, though it were shallow, had a long continuance; whereby they of the vale[120] that were not drowned, perished for want of food and other things necessary. So as[121] marvel you not at the thin population of America, nor at the rudeness and ignorance of the people; for you must account your inhabitants of America as a young people; younger a thousand years, at the least, than the rest of the world; for that there was so much time between the universal flood and their particular inundation. For the poor remnant of human seed which remained in their

---

[116]*Entrapped*

[117]*Timaeus* 25c6

[118]*Not* the universal flood survived by Noah

[119]*In* Plato's account the Athenian warriors were destroyed by earthquakes and floods that caused the whole island of Atlantis to be swallowed by the sea. The war no longer lived in Athenian memory because Athens, along with the rest of the world except for the Egyptians, was subject to various destructions of mankind caused by fire, water, and other means; *Timaeus* 21d2–23d1.

[120]*Valley*

[121]*Therefore*

mountains peopled the country again slowly, by little and little; and being simple and savage people, (not like Noah and his sons, which was the chief family of the earth,) they were not able to leave letters, arts, and civility to their posterity; and having likewise in their mountainous habitations been used (in respect[122] of the extreme cold of those regions) to clothe themselves with the skins of tigers, bears, and great hairy goats, that they have in those parts; when after they came down into the valley, and found the intolerable heats which are there, and knew no means of lighter apparel, they were forced to begin the custom of going naked, which continueth at this day. Only they take great pride and delight in the feathers of birds, and this also they took from those their ancestors of the mountains, who were invited unto it by the infinite flights of birds that came up to the high grounds, while the waters stood below. So you see, by this main accident of time, we lost our traffic with the Americans, with whom of all others, in regard[123] they lay nearest to us, we had most commerce. As for the other parts of the world, it is most manifest that in the ages following (whether it were in respect of wars, or by a natural revolution of time,[124]) navigation did every where greatly decay; and specially far voyages (the rather by the use of galleys, and such vessels as could hardly brook[125] the ocean,) were altogether left and omitted.[126] So then, that part of intercourse which could be from other nations to sail to us, you see how it hath long since ceased; except it were by some rare accident, as this of yours. But now of the cessation of that other part of intercourse, which might be by our sailing to other nations, I must yield you some other cause. For I cannot say (if I shall say truly,) but our shipping, for number, strength, mariners, pilots, and all things that appertain to navigation, is as great as ever:

[122]*Because*

[123]*Since*

[124]*The* natural course of time

[125]*Tolerate*

[126]*As* navigation decayed, "far voyages" came to mean those facilitated by relatively short-range vessels.

and therefore why we should sit at home, I shall now give you an account by itself: and it will draw nearer[127] to give you satisfaction to your principal question.

"There reigned in this island, about nineteen hundred years ago,[128] a King, whose memory of all others we most adore; not superstitiously, but as a divine instrument, though a mortal man; his name was Solamona: and we esteem him as the lawgiver of our nation. This king had a *large heart*,[129] inscrutable for good;[130] and was wholly bent to make his kingdom and people happy. He therefore, taking into consideration how sufficient and substantive[131] this land was to maintain itself without any aid at all of the foreigner; being five thousand six hundred miles in circuit, and of rare fertility of soil in the greatest part thereof; and finding also the shipping of this country might be plentifully set on work, both by fishing and by transportations from port to port, and likewise by sailing unto some small islands that are not far from us, and are under the crown and laws of this state; and recalling into his memory the happy and flourishing estate[132] wherein this land then was, so as it might be a thousand ways altered to the worse, but scarce any one way to the better; thought nothing wanted to his noble and heroical intentions, but[133] only (as far as human foresight might reach) to give perpetuity to that which was in his time so happily established. Therefore amongst his other fundamental laws of this kingdom, he did ordain the interdicts[134] and prohibitions which we have touching entrance of strangers; which at that time (though it was after the calamity of America) was frequent; doubting[135] novelties, and commixture of manners. It is true, the like law against the admission of strangers without licence

[127]*Come* closer to giving
[128]*About* 288 B.C.
[129]*Said* of the Biblical Solomon, 1 Kings 4: 29 (Heb. 5: 9), in reference to his wisdom
[130]*Unfathomably* good, see Prov. 25: 3
[131]*Independent*
[132]*Condition*
[133]*His* noble and heroic intentions needed only
[134]*An* act of forbidding peremptorily
[135]*Fearing*

is an ancient law in the kingdom of China, and yet continued in use. But there it is a poor thing; and hath made them a curious,[136] ignorant, fearful, foolish nation. But our lawgiver made his law of another temper.[137] For first, he hath preserved all points of humanity, in taking order[138] and making provision for the relief of strangers distressed; whereof you have tasted." At which speech (as reason was) we all rose up, and bowed ourselves. He went on. "That king also, still desiring to join humanity and policy together; and thinking it against humanity to detain strangers here against their wills, and against policy that they should return and discover[139] their knowledge of this estate,[140] he took this course: he did ordain that of the strangers that should be permitted to land, as many (at all times) might depart as would; but as many as would stay should have very good conditions and means to live from the state. Wherein he saw so far, that now in so many ages since the prohibition, we have memory not of one ship that ever returned; and but of thirteen persons only, at several[141] times, that chose to return in our bottoms. What those few that returned may have reported abroad I know not. But you must think, whatsoever they have said could be taken where they came but for a dream. Now for our travelling from hence[142] into parts abroad, our Lawgiver thought fit altogether to restrain it. So is it not in China. For the Chineses sail where they will or can; which sheweth that their law of keeping out strangers is a law of pusillanimity and fear. But this restraint of ours hath one only exception, which is admirable; preserving the good which cometh by communicating with strangers, and avoiding the hurt; and I will now open it to you. And here I shall seem a little to digress, but you will by and by find it pertinent. Ye shall understand (my dear friends)

[136]*Anxious,* solicitous
[137]*Character*
[138]*Managing*
[139]*Reveal*
[140]*Kingdom,* commonwealth
[141]*Different*
[142]*Here*

that amongst the excellent acts of that king, one above all hath the preeminence. It was the erection and institution of an Order or Society which we call *Salomon's House;* the noblest foundation (as we think) that ever was upon the earth; and the lanthorn[143] of this kingdom. It is dedicated to the study of the Works and Creatures of God. Some think it beareth the founder's name a little corrupted, as if it should be Solamona's House. But the records write it as it is spoken. So as I take it to be denominate[144] of the King of the Hebrews, which is famous with you, and no stranger to us. For we have some parts of his works which with you are lost; namely, that Natural History which he wrote, of all plants, from the *cedar of Libanus*[145] to the *moss that groweth out of the wall,* and of all *things that have life and motion.*[146] This maketh me think that our king, finding himself to symbolize in many things with[147] that king of the Hebrews (which lived many years before him), honoured him with the title of this foundation. And I am the rather induced to be of this opinion, for that I find in ancient records this Order or Society is sometimes called Salomon's House and sometimes the College of the Six Days' Works; whereby I am satisfied that our excellent king had learned from the Hebrews that God had created the world and all that therein is within six days; and therefore he instituting that House for the finding out of the true nature of all things, (whereby God might have the more glory in the workmanship of them, and men the more fruit in the use of them,) did give it also that second name. But now to come to our present purpose. When the king had forbidden to all his people navigation into any part that was not under his crown, he made nevertheless this ordinance; That every twelve years there should be set forth out of this kingdom two ships, appointed to several[148] voyages; That in either[149] of these ships there should be

[143]*Light*
[144]*Named* for
[145]*Lebanon*
[146]*See* I Kings 4: 33 (Heb. 5: 13).
[147]*To* resemble in many things
[148]*Different*
[149]*Each*

a mission of three of the Fellows or Brethren of Salomon's House; whose errand was only to give us knowledge of the affairs and state of those countries to which they were designed,[150] and especially of the sciences, arts, manufactures, and inventions of all the world; and withal[151] to bring unto us books, instruments, and patterns in every kind; That the ships, after they had landed the brethren, should return; and that the brethren should stay abroad till the new mission. These ships are not otherwise fraught,[152] than with store of victuals, and good quantity of treasure to remain with the brethren, for the buying of such things and rewarding of such persons as they should think fit. Now for me to tell you how the vulgar sort[153] of mariners are contained[154] from being discovered at land; and how they that must be put on shore for any time, colour[155] themselves under the names of other nations; and to what places these voyages have been designed; and what places of *rendez-vous*[156] are appointed for the new missions; and the like circumstances of the practique;[157] I may not do it: neither is it much to your desire. But thus you see we maintain a trade, not for gold, silver, or jewels; nor for silks; nor for spices; nor any other commodity of matter; but only for God's first creature, which was *Light:* to have *light*[158] (I say) of the growth of all parts of the world." And when he had said this, he was silent; and so were we all. For indeed we were all astonished to hear so strange things so probably[159] told. And he, perceiving that we were willing to say somewhat[160] but had it not ready, in great courtesy took us off,[161] and descended[162] to ask

[150]*Bound* for
[151]*In* addition
[152]*Laden*
[153]*Common*
[154]*Kept*
[155]*Misrepresent,* disguise
[156]*Meeting*
[157]*Actual* practice
[158]*Knowledge*
[159]*Plausibly*
[160]*Something*
[161]*Relieved* us
[162]*Lowered* himself

us questions of our voyage and fortunes; and in the end concluded, that we might do well to think with ourselves what time of stay we would demand of the state; and bade us not to scant[163] ourselves; for he would procure such time as we desired. Whereupon we all rose up, and presented ourselves to kiss the skirt of his tippet; but he would not suffer us;[164] and so took his leave. But when it came once amongst our people[165] that the state used to offer conditions[166] to strangers that would stay, we had work enough to get any of our men to look to our ship, and to keep them from going presently to the governor to crave[167] conditions. But with much ado we refrained them, till we might agree what course to take.

We took ourselves now for free men, seeing there was no danger of our utter perdition; and lived most joyfully, going abroad and seeing what was to be seen in the city and places adjacent within our tedder;[168] and obtaining acquaintance with many of the city, not of the meanest quality;[169] at whose hands we found such humanity, and such a freedom and desire to take strangers as it were into their bosom, as was enough to make us forget all that was dear to us in our own countries: and continually we met with many things right worthy of observation and relation; as indeed, if there be a mirror[170] in the world worthy to hold men's eyes, it is that country. One day there were two of our company bidden to a Feast of the Family, as they call it. A most natural, pious, and reverend custom it is, shewing that nation to be compounded of all goodness. This is the manner of it. It is granted to any man that shall live to see thirty persons descended of his body alive together, and all above three years old, to make this feast; which is done at the cost of the state. The Father of the Family, whom they call

[163]*Stint*
[164]*Allow* us to do it
[165]*As* soon as our people heard
[166]*Good* circumstances
[167]*To* beg for
[168]*Tether*
[169]*Poorest*, lowest class
[170]*Model* of excellence

the *Tirsan*,[171] two days before the feast, taketh to him three of such friends as he liketh to choose; and is assisted also by the governor of the city or place where the feast is celebrated; and all the persons of the family, of both sexes, are summoned to attend him. These two days the Tirsan sitteth in consultation concerning the good estate[172] of the family. There, if there be any discord or suits beween any of the family, they are compounded and appeased. There, if any of the family be distressed[173] or decayed,[174] order is taken for their relief and competent means to live. There, if any be subject to vice, or take ill courses,[175] they are reproved and censured. So likewise direction is given touching marriages, and the courses of life which any of them should take, with divers other the like orders and advices. The governor assisteth, to the end to put in execution by his public authority the decrees and orders of the Tirsan, if they should be disobeyed; though that seldom needeth; such reverence and obedience they give to the order of nature. The Tirsan doth also then ever choose one man from amongst his sons, to live in house with him: who is called ever after the Son of the Vine. The reason will hereafter appear. On the feast-day, the Father or Tirsan cometh forth after divine service into a large room where the feast is celebrated; which room hath an half-pace[176] at the upper end. Against the wall, in the middle of the half-pace, is a chair placed for him, with a table and carpet before it. Over the chair is a state,[177] made round or oval, and it is of ivy; an ivy somewhat whiter than ours, like the leaf of a silver asp,[178] but more shining; for it is green all winter. And the state is curiously wrought with silver and silk of divers colours, broiding[179] or binding in the ivy; and is ever of the work of some

[171]*Tirsan* is a Persian word (tarsān) meaning timid or fearful.
[172]*Condition*
[173]*In* difficult financial condition
[174]*Behind* in paying rent
[175]*Act* unwisely
[176]*Dais*
[177]*Canopy*
[178]*Aspen,* poplar
[179]*Interweaving*

of the daughters of the family; and veiled over at the top with a fine net of silk and silver. But the substance of it is true ivy; whereof, after it is taken down, the friends of the family are desirous to have some leaf or sprig to keep. The Tirsan cometh forth with all his generation or lineage, the males before him, and the females following him; and if there be a mother from whose body the whole lineage is descended, there is a traverse[180] placed in a loft above on the right hand of the chair, with a privy[181] door, and a carved window of glass, leaded with gold and blue; where she sitteth, but is not seen. When the Tirsan is come forth, he sitteth down in the chair; and all the lineage place themselves against the wall, both at his back and upon the return[182] of the half-pace, in order of their years without difference of sex; and stand upon their feet. When he is set; the room being always full of company, but well kept and without disorder; after some pause there cometh in from the lower end of the room a *Taratan* (which is as much as an herald) and on either side of him two young lads; whereof one carrieth a scroll of their shining yellow parchment; and the other a cluster of grapes of gold, with a long foot or stalk. The herald and children are clothed with mantles of sea-water green sattin; but the herald's mantle is streamed[183] with gold, and hath a train. Then the herald with three curtesies,[184] or rather inclinations,[185] cometh up as far as the half-pace; and there first taketh into his hand the scroll. This scroll is the King's Charter, containing gift of revenew,[186] and many privileges, exemptions, and points of honour, granted to the Father of the Family; and is ever styled and directed, *To such an one our well-beloved friend and creditor:* which is a title proper only to this case. For they say the king is debtor to no man, but for propagation of his sub-

---

[180]*Screened* compartment
[181]*Concealed*
[182]*Side*
[183]*Striped*
[184]*Bows*
[185]*Bowings* or nods of the head
[186]*Income*

jects.[187] The seal set to the king's charter is the king's image, imbossed or moulded in gold; and though such charters be expedited of course,[188] and as of right, yet they are varied by discretion, according to the number and dignity of the family. This charter the herald readeth aloud; and while it is read, the father or Tirsan standeth up, supported by two of his sons, such as he chooseth. Then the herald mounteth the half-pace, and delivereth the charter into his hand: and with that there is an acclamation by all that are present in their language, which is thus much:[189] *Happy are the people of Bensalem.* Then the herald taketh into his hand from the other child the cluster of grapes, which is of gold, both the stalk and the grapes. But the grapes are daintily enamelled; and if the males of the family be the greater number, the grapes are enamelled purple, with a little sun set on the top; if the females, then they are enamelled into a greenish yellow, with a crescent on the top. The grapes are in number as many as there are descendants of the family. This golden cluster the herald delivereth also to the Tirsan; who presently delivereth it over to that son that he had formerly chosen to be in house with him: who beareth it before his father as an ensign[190] of honour when he goeth in public, ever after; and is thereupon called the Son of the Vine. After this ceremony ended, the father or Tirsan retireth; and after some time cometh forth again to dinner, where he sitteth alone under the state, as before; and none of his descendants sit with him, of what degree or dignity soever, except he hap[191] to be of Salomon's House. He is served only by his own children, such as are male; who perform unto him all service of the table upon the knee; and the women only stand about him, leaning against the wall. The room below the half-pace hath tables on the sides for the guests that are bidden; who are served with great and comely order; and towards the end of dinner (which in

[187]*The* only thing the king needs is the propagation of his subjects.
[188]*Officially* issued or granted
[189]*Which* means roughly
[190]*Badge*
[191]*Happen*

the greatest feasts with them lasteth never above an hour and an half) there is an hymn sung, varied according to the invention of him that composeth it, (for they have excellent poesy,) but the subject of it is (always) the praises of Adam and Noah and Abraham; whereof the former two peopled the world, and the last was the Father of the Faithful: concluding ever with a thanksgiving for the nativity of our Saviour, in whose birth the births of all are only blessed. Dinner being done, the Tirsan retireth again; and having withdrawn himself alone into a place where he maketh some private prayers, he cometh forth the third time, to give the blessing; with all his descendants, who stand about him as at the first. Then he calleth them forth by one and by one, by name, as he pleaseth, though seldom the order of age be inverted. The person that is called (the table being before removed) kneeleth down before the chair, and the father layeth his hand upon his head, or her head, and giveth the blessing in these words: *Son of Bensalem, (or Daughter of Bensalem,) thy father saith it; the man by whom thou hast breath and life speaketh the word; The blessing of the everlasting Father, the Prince of Peace, and the Holy Dove*[192] *be upon thee, and make the days of thy pilgrimage good and many.*[193] This he saith to every of them; and that done, if there be any of his sons of eminent merit and virtue, (so they be not above two,)[194] he calleth for them again; and saith, laying his arm over their shoulders, they standing; *Sons, it is well ye are born, give God the praise, and persevere to the end.* And withal delivereth to either of them a jewel, made in the figure of an ear of wheat, which they ever after wear in the front of their turban or hat. This done, they fall to music and dances, and other recreations, after their manner, for the rest of the day. This is the full order of that feast.

By that time six or seven days were spent, I was fallen into strait[195] acquaintance with a merchant of that city,

[192]*Holy Spirit*
[193]*Genesis* 47: 9
[194]*Only* the two best
[195]*Close*

whose name was Joabin.[196] He was a Jew, and circum-
cised: for they have some few stirps of Jews yet remaining among them, whom they leave to their own
religion. Which they may the better do, because they
are of a far differing disposition from the Jews in other
parts. For whereas they hate the name of Christ, and
have a secret inbred rancour against the people amongst
whom they live: these (contrariwise) give unto our Saviour many high attributes, and love the nation of Bensalem extremely. Surely this man of whom I speak would
ever acknowledge that Christ was born of a Virgin, and
that he was more than a man; and he would tell how
God made him ruler of the Seraphims[197] which guard
his throne; and they call him also the *Milken Way*,[198] and
the *Eliah* of the *Messiah*;[199] and many other high names;
which though they be inferior to his divine Majesty, yet
they are far from the language of other Jews. And for
the country of Bensalem, this man would make no end
of commending it: being desirous, by tradition among
the Jews there, to have it believed that the people thereof
were of the generations of Abraham, by another son,
whom they call Nachoran;[200] and that Moses by a secret
cabala[201] ordained the laws of Bensalem which they now
use; and that when the Messiah should come, and sit in
his throne at Hierusalem,[202] the king of Bensalem should
sit at his feet, whereas other kings should keep a great
distance. But yet setting aside these Jewish dreams, the
man was a wise man, and learned, and of great policy,[203]
and excellently seen[204] in the laws and customs of that

[196]*Plural form of Joab*, i.e., "Joabs." Joab was David's nephew and
an important captain of David's army. See I Samuel 26: 6; II Samuel
2; 3; 8: 16; 10; 11–14; 17: 25; 18–20; 23–24; I Kings 1–2; 11; I Chronicles 2: 16; 11; 18: 15; 19–21; 26: 28; 27.
[197]*Six-winged* angels
[198]*A* way leading to heaven
[199]*Prophetic* forerunner of the Messiah. See Malachi 4: 5 (Heb. 3:
23); Matt. 16: 4; 17: 10.
[200]*Genesis* 11: 22–27; 22: 20–23; 24: 15, 47; 29: 5; I Chronicles 1:
26; Luke 3: 34.
[201]*Doctrine*
[202]*Jerusalem*
[203]*Political* wisdom
[204]*Well* versed

nation. Amongst other discourses, one day I told him I was much affected[205] with the relation I had from some of the company, of their custom in holding the Feast of the Family; for that (methought) I had never heard of a solemnity wherein nature did so much preside. And because propagation of families proceedeth from the nuptial copulation, I desired to know of him what laws and customs they had concerning marriage; and whether they kept marriage well; and whether they were tied to one wife? For that where population is so much affected,[206] and such as with them it seemed to be, there is commonly permission of plurality of wives. To this he said, "You have reason for to commend that excellent institution of the Feast of the Family. And indeed we have experience, that those families that are partakers of the blessing of that feast do flourish and prosper ever after in an extraordinary manner. But hear me now, and I will tell you what I know. You shall understand that there is not under the heavens so chaste a nation as this of Bensalem; nor so free from all pollution or foulness. It is the virgin of the world. I remember I have read in one of your European books,[207] of an holy hermit amongst you that desired to see the Spirit of Fornication; and there appeared to him a little foul ugly Æthiop.[208] But if he had desired to see the Spirit of Chastity of Bensalem, it would have appeared to him in the likeness of a fair beautiful Cherubin. For there is nothing amongst mortal men more fair and admirable, than the chaste minds of this people. Know therefore, that with them there are no stews,[209] no dissolute houses, no courtesans, nor any thing of that kind. Nay they wonder (with detestation) at you in Europe, which permit such things. They say ye have put marriage out of office: for marriage is ordained a remedy for unlawful concupiscence; and natural concupiscence seemeth as a spur to marriage. But when men have at hand a remedy more agreeable

[205]*Very* impressed by
[206]*Desired*
[207]*Sintram* by La Motte Fouque
[208]*Blackamoor*, Klein Meister in *Sintram*
[209]*Brothels*

to their corrupt will, marriage is almost expulsed. And
therefore there are with you seen infinite men that marry
not, but chuse rather a libertine and impure single life,
than to be yoked in marriage; and many that do marry,
marry late, when the prime and strength of their years
is past. And when they do marry, what is marriage to
them but a very bargain; wherein is sought alliance, or
portion,[210] or reputation, with some desire (almost indifferent) of issue;[211] and not the faithful nuptial union
of man and wife, that was first instituted. Neither is it
possible that those that have cast away so basely so much
of their strength, should greatly esteem children, (being
of the same matter,) as chaste men do. So likewise during
marriage, is the case much amended,[212] as it ought to
be if those things were tolerated only for necessity? No,
but they remain still as a very affront to marriage. The
haunting of those dissolute places, or resort to courtesans, are no more punished in married men than in bach-
elors. And the depraved custom of change,[213] and the
delight in meretricious embracements, (where sin is
turned into art,) maketh marriage a dull thing, and a
kind of imposition or tax. They hear you defend these
things, as done to avoid greater evils; as advoutries,[214]
deflouring of virgins, unnatural lust, and the like. But
they say this is a preposterous wisdom; and they call it
Lot's offer,[215] who to save his guests from abusing, offered his daughters: nay they say farther that there is
little gained in this; for that the same vices and appetites
do still remain and abound; unlawful lust being like a
furnace, that if you stop the flames altogether, it will
quench; but if you give it any vent, it will rage. As for
masculine love, they have no touch of it; and yet there
are not so faithful and inviolate friendships in the world
again as are there; and to speak generally, (as I said
before,) I have not read of any such chastity in any

---

[210]*Dowry*
[211]*Only* a faint desire for children
[212]*Is* the situation different
[213]*Alteration*
[214]*Adulteries*
[215]*See* Genesis 19: 1–11.

people as theirs. And their usual saying is, *That whosoever is unchaste cannot reverence himself*; and they say, *That the reverence of a man's self is, next religion, the chiefest bridle of all vices."* And when he had said this, the good Jew paused a little; whereupon I, far more willing to hear him speak on than to speak myself, yet thinking it decent that upon his pause of speech I should not be altogether silent, said only this; "That I would say to him, as the widow of Sarepta said to Elias;[216] that he was come to bring to memory our sins; and that I confess the righteousness of Bensalem was greater than the righteousness of Europe." At which speech he bowed his head, and went on in this manner: "They have also many wise and excellent laws touching marriage. They allow no polygamy. They have ordained that none do intermarry or contract, until a month be passed from their first interview. Marriage without consent of parents they do not make void, but they mulct[217] it in the inheritors: for the children of such marriages are not admitted to inherit above a third part of their parents' inheritance. I have read in a book of one of your men, of a Feigned Commonwealth, where the married couple are permitted, before they contract, to see one another naked.[218] This they dislike; for they think it a scorn to give a refusal after so familiar knowledge: but because of many hidden defects in men and women's bodies, they have a more civil way; for they have near every town a couple of pools, (which they call *Adam and Eve's pools,)* where it is permitted to one of the friends of the man, and another of the friends of the woman, to see them severally[219] bathe naked."[220]

And as we were thus in conference, there came one that seemed to be a messenger, in a rich huke,[221] that spake with the Jew: whereupon he turned to me and

[216]*See* I Kings 17.
[217]*Punish* by a fine
[218]*Thomas* More, *Utopia*; "Their Marriage Customs"; Plato, *Laws* 771e.
[219]*Separately*
[220]*Cf.* II Samuel 11: 2–27.
[221]*Hooded* cape

said; "You will pardon me, for I am commanded away in haste." The next morning he came to me again, joyful as it seemed, and said, "There is word come to the governor of the city, that one of the Fathers of Salomon's House will be here this day seven-night:[222] we have seen none of them this dozen years. His coming is in state;[223] but the cause of his coming is secret. I will provide you and your fellows of a good standing to see his entry." I thanked him, and told him, "I was most glad of the news." The day being come, he made his entry. He was a man of middle stature and age, comely of person, and had an aspect as if he pitied men. He was clothed in a robe of fine black cloth, with wide sleeves and a cape. His under garment was of excellent white linen down to the foot, girt with a girdle of the same; and a sindon or tippet of the same about his neck. He had gloves that were curious,[224] and set with stone; and shoes of peach-coloured velvet. His neck was bare to the shoulders. His hat was like a helmet, or Spanish Montera;[225] and his locks curled below it decently:[226] they were of colour brown. His beard was cut round, and of the same colour with his hair, somewhat lighter. He was carried in a rich chariot without wheels, litter-wise; with two horses at either end, richly trapped[227] in blue velvet embroidered; and two footmen on each side in the like attire. The chariot was all of cedar, gilt, and adorned with crystal; save that the fore-end had pannels[228] of sapphires, set in borders of gold, and the hinder-end the like of emeralds of the Peru colour.[229] There was also a sun of gold, radiant, upon the top, in the midst; and on the top before, a small cherub of gold, with wings displayed. The chariot was covered with cloth of gold tissued[230] upon blue. He had before him fifty attendants, young

[222]*A* week from today
[223]*With* great pomp and solemnity
[224]*Elaborate,* carefully made
[225]*A* round cap
[226]*Becomingly*
[227]*Adorned*
[228]*Panels*
[229]*True* emeralds
[230]*Interwoven*

men all, in white sattin loose coats to the mid-leg; and
stockings of white silk; and shoes of blue velvet; and
hats of blue velvet; with fine plumes of divers colours,
set round like hat-bands. Next before the chariot went
two men, bare-headed, in linen garments down to the
foot, girt, and shoes of blue velvet; who carried the one
a crosier,[231] the other a pastoral staff[232] like a sheep-
hook; neither of them of metal, but the crosier of balm-
wood,[233] the pastoral staff of cedar. Horsemen he had
none, neither before nor behind his chariot: as it seemeth, to avoid all tumult and trouble. Behind his chariot
went all the officers and principals of the Companies of
the City.[234] He sat alone, upon cushions of a kind of
excellent plush, blue; and under his foot curious carpets
of silk of divers colours, like the Persian, but far finer.
He held up his bare hand as he went, as blessing the
people, but in silence. The street was wonderfully well
kept: so that there was never any army had their men
stand in better battle-array, than the people stood. The
windows likewise were not crowded, but every one stood
in them as if they had been placed. When the shew was
past, the Jew said to me; "I shall not be able to attend
you as I would, in regard of some charge the city hath
laid upon me, for the entertaining of this great person."
Three days after, the Jew came to me again, and said;
"Ye are happy men; for the Father of Salomon's House
taketh knowledge of your being here, and commanded
me to tell you that he will admit all your company to
his presence, and have private conference with one of
you that ye shall choose: and for this hath appointed
the next day after to-morrow. And because he meaneth
to give you his blessing, he hath appointed it in the
forenoon." We came at our day and hour, and I was
chosen by my fellows for the private access. We found
him in a fair chamber, richly hanged,[235] and carpeted

[231]***Cross*** of an archbishop
[232]***Staff*** borne by a bishop
[233]***Balsam***
[234]***Trade*** guilds
[235]***Adorned*** with tapestry

under foot, without any degrees[236] to the state.[237] He was set upon a low throne richly adorned, and a rich cloth of state[238] over his head, of blue satin embroidered. He was alone, save that he had two pages of honour, on either hand one, finely attired in white. His undergarments were the like that we saw him wear in the chariot; but instead of his gown, he had on him a mantle with a cape, of the same fine black, fastened about him. When we came in, as we were taught, we bowed low at our first entrance; and when we were come near his chair, he stood up, holding forth his hand ungloved, and in posture of blessing; and we every one of us stooped down, and kissed the hem of his tippet. That done, the rest departed, and I remained. Then he warned the pages forth of the room, and caused me to sit down beside him, and spake to me thus in the Spanish tongue:

"God bless thee, my son; I will give thee the greatest jewel I have. For I will impart unto thee, for the love of God and men, a relation of the true state of Salomon's House. Son, to make you know the true state of Salomon's House, I will keep this order. First, I will set forth unto you the end of our foundation. Secondly, the preparations and instruments we have for our works. Thirdly, the several employments and functions whereto our fellows are assigned. And fourthly, the ordinances and rites which we observe.

"The End of our Foundation is the knowledge of Causes, and secret motions of things; and the enlarging of the bounds of Human Empire, to the effecting of all things possible.

"The Preparations and Instruments are these. We have large and deep caves of several depths: the deepest are sunk six hundred fathom; and some of them are digged and made under great hills and mountains: so that if you reckon together the depth of the hill and the depth of the cave, they are (some of them) above three miles deep. For we find that the depth of a hill, and the depth

[236]***Steps***
[237]***Throne***
[238]***Official*** cloth

of a cave from the flat, is the same thing; both remote alike from the sun and heaven's beams, and from the open air.[239] These caves we call the Lower Region. And we use them for all coagulations, indurations,[240] refrigerations, and conservations of bodies. We use them likewise for the imitation of natural mines;[241] and the producing also of new artificial metals, by compositions and materials which we use, and lay there for many years. We use them[242] also sometimes, (which may seem strange,) for curing of some diseases, and for prolongation of life in some hermits that choose to live there, well accommodated of[243] all things necessary; and indeed live very long; by whom also we learn many things.

"We have burials in several earths, where we put divers cements, as the Chineses do their porcellain. But we have them in greater variety, and some of them more fine. We have also great variety of composts, and soils, for the making of the earth fruitful.

"We have high towers; the highest about half a mile in height; and some of them likewise set upon high mountains; so that the vantage[244] of the hill with the tower is in the highest of them three miles at least. And these places we call the Upper Region: accounting the air between the high places and the low, as a Middle Region. We use these towers, according to their several heights and situations, for insolation,[245] refrigeration, conservation; and for the view of divers meteors;[246] as winds, rain, snow, hail; and some of the fiery meteors[247] also. And upon them, in some places, are dwellings of hermits, whom we visit sometimes, and instruct what to observe.

"We have great lakes both salt and fresh, whereof we have use for the fish and fowl. We use them also for

[239]*Thus,* caves dug downward under hills are especially deep.
[240]*Hardenings*
[241]*Mineral* veins
[242]*The* caves
[243]*Well* provided with
[244]*Total* height
[245]*Exposure* to the sun
[246]*Atmospheric* phenomena
[247]*Shooting* stars and lightning

burials of some natural bodies: for we find a difference in things buried in earth or in air below the earth, and things buried in water. We have also pools, of which some do strain fresh water out of salt; and others by art do turn fresh water into salt. We have also some rocks in the midst of the sea, and some bays upon the shore, for some works wherein is required the air and vapour of the sea. We have likewise violent streams and cataracts,[248] which serve us for many motions: and likewise engines for multiplying and enforcing of winds, to set also on going[249] divers motions.

"We have also a number of artificial wells and fountains, made in imitation of the natural sources and baths; as tincted upon[250] vitriol,[251] sulphur, steel, brass, lead, nitre,[252] and other minerals. And again we have little wells for infusions[253] of many things, where the waters take the virtue[254] quicker and better than in vessels or basons. And amongst them we have a water which we call Water of Paradise, being, by that we do to it, made very sovereign[255] for health, and prolongation of life.

"We have also great and spacious houses, where we imitate and demonstrate meteors; as snow, hail, rain, some artificial rains of bodies and not of water, thunders, lightnings; also generations of bodies in air; as frogs, flies, and divers others.

"We have also certain chambers, which we call Chambers of Health, where we qualify[256] the air as we think good and proper for the cure of divers diseases, and preservation of health.

"We have also fair and large baths, of several mixtures, for the cure of diseases, and the restoring of man's body from arefaction:[257] and others for the confirming[258] of it

[248]*Steep* waterfalls
[249]*To* produce
[250]*Tinged* with
[251]*Sulphate* of metal
[252]*Saltpeter*
[253]*Modifications* by mixing in new elements
[254]*Medicinal* efficacy
[255]*Most* efficacious
[256]*Alter,* change
[257]*Withering*
[258]*Strengthening*

in strength of sinews, vital parts, and the very juice and substance of the body.

"We have also large and various orchards and gardens, wherein we do not so much respect beauty, as variety of ground and soil, proper for divers trees and herbs: and some very spacious, where trees and berries are set whereof we make divers kinds of drinks, besides the vineyards. In these we practise likewise all conclusions[259] of grafting and inoculating,[260] as well of wild-trees as fruit-trees, which produceth many effects. And we make (by art) in the same orchards and gardens, trees and flowers to come[261] earlier or later than their seasons; and to come up and bear more speedily than by their natural course they do. We make them also by art greater much than their nature; and their fruit greater and sweeter and of differing taste, smell, colour, and figure, from their nature. And many of them we so order,[262] as they become of medicinal use.

"We have also means to make divers plants rise by mixtures of earths without seeds; and likewise to make divers new plants, differing from the vulgar;[263] and to make one tree or plant turn into another.

"We have also parks and inclosures of all sorts of beasts and birds, which we use not only for view or rareness,[264] but likewise for dissections and trials; that thereby we may take light[265] what may be wrought upon[266] the body of man. Wherein we find many strange effects; as continuing life in them, though divers parts, which you account vital, be perished and taken forth; resuscitating of some that seem dead in appearance; and the like. We try also all poisons and other medicines upon them, as well of chirurgery[267] as physic.[268] By art likewise, we

[259]*Experiments*
[260]*Budding*
[261]*Germinate*
[262]*Cause* to germinate faster or slower, grow faster, etc.
[263]*Common*
[264]*For* display and preservation of rare species
[265]*Learn*
[266]*Done* to
[267]*Surgery*
[268]*Medicine*

make them greater or taller than their kind is; and contrariwise dwarf them, and stay their growth: we make them more fruitful and bearing than their kind is; and contrariwise barren and not generative. Also we make them differ in colour, shape, activity, many ways. We find means to make commixtures and copulations[269] of different kinds; which have produced many new kinds, and them not barren, as the general opinion is. We make a number of kinds of serpents, worms, flies, fishes, of putrefaction,[270] whereof some are advanced (in effect) to be perfect creatures,[271] like beasts or birds; and have sexes, and do propagate. Neither do we this by chance, but we know beforehand of what matter and commixture what kind of those creatures will arise.

"We have also particular pools, where we make trials upon fishes, as we have said before of beasts and birds.

"We have also places for breed and generation of those kinds of worms and flies which are of special use; such as are with you your silk-worms and bees.

"I will not hold you long with recounting of our brew-houses, bake-houses, and kitchens, where are made divers drinks, breads, and meats, rare and of special effects. Wines we have of grapes; and drinks of other juice of fruits, of grains, and of roots: and of mixtures with honey, sugar, manna, and fruits dried and decocted.[272] Also of the tears[273] or woundings of trees, and of the pulp of canes. And these drinks are of several ages, some to the age or last[274] of forty years. We have drinks also brewed with several herbs, and roots, and spices; yea with several fleshes, and white meats; whereof some of the drinks are such, as they are in effect meat and drink both: so that divers, especially in age,[275] do desire to live with them, with little or no meat or bread. And above all, we strive to have drinks of extreme thin parts,[276] to

[269]***Mixings*** and couplings
[270]***Out*** of decaying matter
[271]***Complex*** enough to reproduce
[272]***Condensed***
[273]***Drops*** of liquid exuded spontaneously
[274]***Duration***
[275]***Various*** people, especially the aged
[276]***Extremely*** fine composition

insinuate[277] into the body, and yet without all[278] biting, sharpness, or fretting;[279] insomuch as some of them put upon the back of your hand will, with a little stay,[280] pass through to the palm, and yet taste mild to the mouth. We have also waters which we ripen[281] in that fashion, as they become nourishing; so that they are indeed excellent drink; and many will use no other. Breads we have of several grains, roots, and kernels: yea and some of flesh and fish dried; with divers kinds of leavenings and seasonings: so that some do extremely move appetites;[282] some do nourish so, as divers[283] do live of them, without any other meat; who live very long. So for meats, we have some of them so beaten and made tender and mortified,[284] yet without all corrupting,[285] as a weak heat of the stomach will turn them into good chylus,[286] as well as a strong heat would meat otherwise prepared. We have some meats also and breads and drinks, which taken by men enable them to fast long after; and some other, that used make the very flesh of men's bodies sensibly[287] more hard and tough, and their strength far greater than otherwise it would be.

"We have dispensatories,[288] or shops of medicines. Wherein you may easily think, if we have such variety of plants and living creatures more than you have in Europe, (for we know what you have,) the simples,[289] drugs, and ingredients of medicines, must likewise be in so much the greater variety. We have them likewise of divers ages, and long fermentations. And for their preparations, we have not only all manner of exquisite

[277]*Penetrate* subtly
[278]*Any*
[279]*Corrosion*
[280]*If* allowed to remain there a short time
[281]*Develop,* perfect
[282]*Are* very delectable
[283]*Some* people
[284]*Tenderized* by hanging
[285]*Any* rotting
[286]*Chyle,* fluid in the intestines just before absorption
[287]*To* an appreciable degree
[288]*Dispensaries*
[289]*Unmixed,* basic substance of medicine, herbs

distillations and separations, and especially by gentle heats and percolations through divers strainers, yea and substances; but also exact forms of composition, whereby they incorporate[290] almost, as they were natural simples.[291]

"We have also divers mechanical arts, which you have not; and stuffs made by them; as papers, linen, silks, tissues; dainty works of feathers of wonderful lustre; excellent dyes, and many others; and shops likewise, as well for such as are not brought into vulgar use amongst us as for those that are. For you must know that of the things before recited, many of them are grown into use throughout the kingdom; but yet if they did flow from our invention, we have of them also for patterns and principals.[292]

"We have also furnaces of great diversities, and that keep great diversity of heats; fierce and quick; strong and constant; soft and mild; blown, quiet; dry, moist; and the like. But above all, we have heats in imitation of the sun's and heavenly bodies' heats, that pass divers inequalities[293] and (as it were) orbs,[294] progresses,[295] and returns,[296] whereby we produce admirable effects. Besides, we have heats of dungs, and of bellies and maws of living creatures, and of their bloods and bodies; and of hays and herbs laid up moist; of lime unquenched;[297] and such like. Instruments also which generate heat only by motion. And farther, places for strong insolations; and again, places under the earth, which by nature or art yield heat. These divers heats we use, as the nature of the operation which we intend requireth.

"We have also perspective-houses,[298] where we make demonstrations of all lights and radiations; and of all colours; and out of things uncoloured and transparent, we can represent unto you all several colours; not in

[290]*Unite* or blend into new substances
[291]*Medicines* composed of one ingredient
[292]*Preserve* them only as examples or originals
[293]*Changes* in intensity
[294]*Planes*
[295]*Forward* courses
[296]*Return* courses
[297]*Unslaked*
[298]*Houses* for optical instruments

rain-bows, as it is in gems and prisms, but of themselves single. We represent[299] also all multiplications[300] of light, which we carry to great distance, and make so sharp as to discern small points and lines; also all colorations of light: all delusions and deceits[301] of the sight, in figures, magnitudes, motions, colours: all demonstrations of shadows. We find also divers means, yet unknown to you, of producing of light originally from divers bodies. We procure means of seeing objects afar off; as in the heaven and remote places; and represent things near as afar off, and things afar off as near; making feigned distances. We have also helps for the sight, far above spectacles and glasses in use. We have also glasses and means to see small and minute bodies perfectly and distinctly; as the shapes and colours of small flies and worms, grains and flaws in gems, which cannot otherwise be seen; observations in urine and blood, not otherwise to be seen. We make artificial rain-bows, halos, and circles about light. We represent also all manner of reflexions, refractions, and multiplications of visual beams of objects.

"We have also precious stones of all kinds, many of them of great beauty, and to you unknown; crystals likewise; and glasses of divers kinds; and amongst them some of metals vitrificated,[302] and other materials besides those of which you make glass. Also a number of fossils,[303] and imperfect minerals, which you have not. Likewise loadstones[304] of prodigious virtue;[305] and other rare stones, both natural and artificial.

"We have also sound-houses, where we practise and demonstrate all sounds, and their generation. We have harmonies which you have not, of quarter-sounds,[306] and lesser slides[307] of sounds. Divers instruments of music likewise to you unknown, some sweeter than any you

[299]*Produce*
[300]*Intensifications*
[301]*Illusions*
[302]*Converted* into a glass-like substance
[303]*Any* rock or mineral dug out of the earth
[304]*Magnets*
[305]*Power*
[306]*Quarter-notes*
[307]*Graces*

have; together with bells and rings[308] that are dainty and sweet. We represent small sounds as great and deep; likewise great sounds extenuate[309] and sharp; we make divers tremblings and warblings of sounds, which in their original are entire.[310] We represent and imitate all articulate sounds and letters, and the voices and notes of beasts and birds. We have certain helps which set to the ear do further the hearing greatly. We have also divers strange and artificial echos, reflecting the voice many times, and as it were tossing it: and some that give back the voice louder than it came; some shriller, and some deeper; yea, some rendering the voice differing in the letters or articulate sound from that they receive. We have also means to convey sounds in trunks[311] and pipes, in strange lines[312] and distances.

"We have also perfume-houses; wherewith we join also practices[313] of taste. We multiply smells, which may seem strange. We imitate smells, making all smells to breathe out of other mixtures than those that give them. We make divers imitations of taste likewise, so that they will deceive any man's taste. And in this house we contain also a confiture-house;[314] where we make all sweet-meats,[315] dry and moist, and divers pleasant wines, milks, broths, and sallets,[316] far in greater variety than you have.

"We have also engine-houses, where are prepared engines and instruments for all sorts of motions. There we imitate and practise to make swifter motions than any you have, either out of your muskets or any engine that you have; and to make them and multiply them more easily, and with small force, by wheels and other means: and to make them stronger, and more violent than yours are; exceeding your greatest cannons and basilisks.[317] We

[308]*Chimes*
[309]*Thin*
[310]*Whole*
[311]*Tubes*
[312]*Directions*
[313]*Studies*
[314]*Confection* house
[315]*Candied* fruits, nuts, etc.
[316]*Salads*
[317]*Large* cannons named for a deadly serpent

represent also ordnance and instruments of war, and engines of all kinds: and likewise new mixtures and compositions of gun-powder, wildfires burning in water, and unquenchable. Also fire-works of all variety both for pleasure and use. We imitate also flights of birds; we have some degrees[318] of flying in the air; we have ships and boats for going under water, and brooking of seas; also swimming-girdles[319] and supporters. We have divers curious[320] clocks, and other like motions of return,[321] and some perpetual motions. We imitate also motions of living creatures, by images[322] of men, beasts, birds, fishes, and serpents. We have also a great number of other various motions, strange for equality, fineness, and subtilty.[323]

"We have also a mathematical house, where are represented all instruments, as well of geometry as astronomy, exquisitely made.

"We have also houses of deceits of the senses; where we represent all manner of feats of juggling, false apparitions, impostures, and illusions; and their fallacies.[324] And surely you will easily believe that we that have so many things truly natural which induce admiration, could in a world of particulars[325] deceive the senses, if we would disguise those things and labour to make them seem more miraculous. But we do hate all impostures and lies: insomuch as[326] we have severely forbidden it to all our fellows, under pain of ignominy and fines, that they do not[327] shew any natural work or thing, adorned or swelling;[328] but only pure as it is, and without all affectation of strangeness.

"These are (my son) the riches of Salomon's House.

[318]*Some* success
[319]*Life* preservers
[320]*Precise*
[321]*Machines* producing cyclical motion
[322]*Robots,* automata
[323]*Extraordinary* for regularity, sharpness, and complexity
[324]*How* they are false
[325]*That* could in the practical world
[326]*So* much that
[327]*An* acceptable double negative in Bacon's time
[328]*Exaggerated*

"For the several employments and offices of our fellows; we have twelve that sail into foreign countries, under the names of other nations, (for our own we conceal;) who bring us the books, and abstracts,[329] and patterns of experiments of all other parts. These we call Merchants of Light.

"We have three that collect the experiments which are in all books. These we call Depredators.[330]

"We have three that collect the experiments of all mechanical arts; and also of liberal sciences;[331] and also of practices which are not brought into arts.[332] These we call Mystery-men.

"We have three that try new experiments, such as themselves think good. These we call Pioners or Miners.

"We have three that draw the experiments of the former four into titles and tables,[333] to give the better light[334] for the drawing of observations and axioms out of them. These we call Compilers.

"We have three that bend themselves,[335] looking into the experiments of their fellows, and cast about[336] how to draw out of them things of use and practice for man's life, and knowledge as well for works as for plain demonstration[337] of causes, means of natural divinations,[338] and the easy and clear discovery of the virtues[339] and parts of bodies. These we call Dowry-men or Benefactors.

"Then after divers meetings and consults[340] of our whole number, to consider of the former labours and collections, we have three that take care, out of them, to direct

[329]*Summaries*
[330]*Those* who plunder or pillage
[331]*Liberal* arts
[332]*Unsystematic* practices
[333]*Those* who classify and list the experiments of the former
[334]*The* better to enable
[335]*Direct* their attention to
[336]*Consider*
[337]*Theoretical* demonstration
[338]*Discovering* and predicting the secrets of nature
[339]*Characteristics,* powers
[340]*Consultations*

new experiments, of a higher light,[341] more penetrating into nature than the former. These we call Lamps.

"We have three others that do execute the experiments so directed, and report them. These we call Inoculators.[342]

"Lastly, we have three that raise the former discoveries by experiments into greater observations, axioms, and aphorisms. These we call Interpreters of Nature.

"We have also, as you must think, novices and apprentices, that the succession of the former employed men do not fail; besides a great number of servants and attendants, men and women. And this we do also: we have consultations, which of the inventions and experiences which we have discovered shall be published, and which not: and take all an oath of secrecy, for the concealing of those which we think fit to keep secret: though some of those we do reveal sometimes to the state, and some not.

"For our ordinances and rites: we have two very long and fair galleries: in one of these we place patterns and samples of all manner of the more rare and excellent inventions: in the other we place the statua's[343] of all principal inventors. There we have the statua of your Columbus, that discovered the West Indies: also the inventor of ships: your monk[344] that was the inventor of ordnance and of gunpowder: the inventor of music: the inventor of letters: the inventor of printing: the inventor of observations of astronomy: the inventor of works in metal: the inventor of glass: the inventor of silk of the worm: the inventor of wine: the inventor of corn and bread: the inventor of sugars: and all these by more certain[345] tradition than you have. Then have we divers inventors of our own, of excellent works; which since you have not seen, it were too long to make descriptions of them; and besides, in the right understanding of those

[341]*Producing* more general knowledge
[342]*Men* who bud trees
[343]*Statues*
[344]*Roger Bacon* or Berthold Schwarz
[345]*More* trustworthy

descriptions you might easily err. For upon every invention of value, we erect a statua to the inventor, and
give him a liberal and honourable reward. These statua's
are some of brass; some of marble and touch-stone;[346]
some of cedar and other special woods gilt and adorned:
some of iron; some of silver; some of gold.

"We have certain hymns and services, which we say
daily, of laud,[347] and thanks to God for his marvellous
works: and forms of prayers, imploring his aid and blessing for the illumination of our labours, and the turning
of them into good and holy uses.

"Lastly, we have circuits or visits of divers principal
cities of the kingdom; where, as it cometh to pass, we
do publish such new profitable inventions as we think
good. And we do also declare natural divinations of diseases, plagues, swarms of hurtful creatures, scarcity,
tempests, earthquakes, great inundations, comets,
temperature[348] of the year, and divers other things; and
we give counsel thereupon what the people shall do for
the prevention and remedy of them."

And when he had said this, he stood up; and I, as I
had been taught, kneeled down; and he laid his right
hand upon my head, and said; "God bless thee, my son,
and God bless this relation which I have made. I give
thee leave to publish it for the good of other nations;
for we here are in God's bosom, a land unknown." And
so he left me; having assigned a value of about two
thousand ducats,[349] for a bounty[350] to me and my fellows.
For they give great largesses[351] where they come upon
all occasions.

[THE REST WAS NOT PERFECTED.]

[346]***Dark*** quartz or jasper
[347]***In*** praise of
[348]***Climate***
[349]***Gold*** coins
[350]***Gift,*** gratuity
[351]***Free*** gifts

bibliography

biographies

Aubrey, John. *Brief Lives*. Edited from the original mss. and with a life of John Aubrey by Oliver Lawson Dick; foreword by Edmund Wilson. Ann Arbor: University of Michigan Press, 1962.

Bowen, Catherine Drinker. *Francis Bacon: The Temper of a Man*. Boston: Little, Brown and Co., 1963.

Church, R. W. *Francis Bacon*. New York, 1884.

Sturt, Mary. *Francis Bacon*. New York: William Morrow and Co., 1932.

bacon's works

The standard, definitive edition of Bacon's works is by Spedding, Ellis and Heath, 14 Vols. (London, 1857–1874). This includes Rawley's *The Life of the Right Honorable Francis Bacon*.

Translations of *Temporis Partus Masculus (The Masculine Birth of Time)*, *Cogitata et Visa (Thoughts and Conclusions)*, and *Redargutio Philosophiarum (The Refutation of Philosophies)* can be found in Benjamin Farrington, *The Philosophy of Francis Bacon*. Liverpool University Press, 1964.

Criticism

Anderson, F. H. *The Philosophy of Francis Bacon*. Chicago: University of Chicago Press, 1948.

Jardine, Lisa. *Francis Bacon: Discovery and the Art of Discourse*. Cambridge: Cambridge University Press, 1974.

Macaulay, T. B. *Critical and Historical Essays.* New York, 1951.

Quinton, Anthony. *Francis Bacon.* Oxford: Oxford University Press, 1980.

Rossi, Paolo. *Francis Bacon: From Magic to Science.* Translated by Sacha Rabinovitch. Chicago: University of Chicago Press, 1968.

Vickers, Brian, ed. *Essential Articles for the Study of Francis Bacon.* Hamden, Ct.: Shoe String Press, 1968.

Weinberger, Jerry. "Science and Rule in Bacon's Utopia: An Introduction to the Reading of the *New Atlantis*," *American Political Science Review* 70 (September 1976): 865–85.

Weinberger, Jerry. *Science, Faith, and Politics: Francis Bacon and the Utopian Roots of the Modern Age.* Ithaca: Cornell University Press, 1985.

White, Howard B. *Peace Among the Willows: The Political Philosophy of Francis Bacon.* The Hague: Martinus Nijhoff, 1968.

Whitney, Charles. *Francis Bacon and Modernity.* New Haven: Yale University Press, 1986.